Ecological Pedagogy, Buddhist Pedagogy, Hermeneutic Pedagogy

Studies in the Postmodern Theory of Education

Shirley R. Steinberg
General Editor

Vol. 452

The Counterpoints series is part of the Peter Lang Education list.
Every volume is peer reviewed and meets
the highest quality standards for content and production.

PETER LANG
New York • Washington, D.C./Baltimore • Bern
Frankfurt • Berlin • Brussels • Vienna • Oxford

JACKIE SEIDEL & DAVID W. JARDINE

Ecological Pedagogy, Buddhist Pedagogy, Hermeneutic Pedagogy

EXPERIMENTS IN A CURRICULUM FOR MIRACLES

PETER LANG
New York • Washington, D.C./Baltimore • Bern
Frankfurt • Berlin • Brussels • Vienna • Oxford

Library of Congress Cataloging-in-Publication Data

Seidel, Jackie.
Ecological pedagogy, Buddhist pedagogy, hermeneutic pedagogy:
experiments in a curriculum for miracles / Jackie Seidel, David W. Jardine.
pages cm. — (Counterpoints: studies in the
postmodern theory of education; vol. 452)
Includes bibliographical references and index.
1. Education—Philosophy. 2. Education—Curricula—Philosophy.
3. Transformative learning. 4. Postmodernism and education.
5. Buddhism—Philosophy. 6. Ecology. 7. Hermeneutics.
I. Jardine, David William. II. Title.
LB14.7.S43 370.1—dc23 2013038196
ISBN 978-1-4331-2253-8 (hardcover)
ISBN 978-1-4331-2252-1 (paperback)
ISBN 978-1-4539-1233-1 (e-book)
ISSN 1058-1634

Bibliographic information published by **Die Deutsche Nationalbibliothek**.
Die Deutsche Nationalbibliothek lists this publication in the "Deutsche
Nationalbibliografie"; detailed bibliographic data is available
on the Internet at http://dnb.d-nb.de/.

Cover illustration: *The Aviator* by Connie Geerts

The paper in this book meets the guidelines for permanence and durability
of the Committee on Production Guidelines for Book Longevity
of the Council of Library Resources.

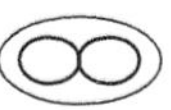

29 Broadway, 18th floor, New York, NY 10006
www.peterlang.com

Printed in the United States of America

Table of Contents

Dedication

This book is dedicated to David G. Smith. Words fail us, again and again, in describing the intimate difference one being can make in the life of another when one lives in the presence of such a boundless heart. We dedicate whatever meager merit may have accrued around us from knowing and loving him to the great and irremediable suffering of all beings, especially those in schools. This book is our thanksgiving.

This book is also dedicated to Peter Rilstone for all the years he has taught us.

Introduction: "We Are Here, We Are Here."

JACKIE SEIDEL & DAVID W. JARDINE

Preamble

In the coulee a pocket of darkness.
Marbled pairs of reflected light,
briefly glow, then shimmer and fade out.
Alone now, they wind through tangles, relentless,
and re-emerge into one.
Call up to the creators; we are here, we are here.
from Judson Innes, "A Pocket Full of Darkness"

We begin with this passage from a poem by Judson Innes (see Chapter 8) because it contains a summoning, a calling, similar in tone to the way that Geshe Lhundub Sopa begins his enormous, multi-volume commentary on Tsong-kha-pa's *The Great Treatise on the Stages of the Path to Enlightenment* (2000, 2002, 2004).

At the very beginning of the 1720 pages of three volumes published thus far (2004, 2005, 2008), Geshe Sopa (2004, p. 1) starts thus:

"So, here we are. Right now, you have a life that is precious and valuable."

So many teachers and students that we have worked with over many years have done beautiful work that calls out, over and over, "we are here, we are here." So many have suffered enormously in the confines of schooling and its bluntings, fears, and shams. It is in recognition of these teachers and these students—our teachers, our students—that we undertake this book.

Again from Geshe Sopa (2004, p. 15), in an early Prologue section entitled "A Pledge to Compose this Work": "Now, why does an author have to make a promise to write? Someone can compose a text without making promise to do so. Here, this pledge has a special purpose: to publically proclaim, 'I will do this.'" Our pledge, one that slowly emerged as this text itself emerged, adds this: this beautiful work between students and teachers, inside and outside of the confines of schooling, *exists*, sometimes in silence and isolation, sometimes in enclaves of refuge and support. Not only, then, "I will do this," but also "this has been done," "this can be done." This is why we saw fit in some of our recent graduate classes to call these gatherings "refuges." This is why we ate together, talked and read and listened and fell silent together. Up against the too often pronounced exhaustion and desperation and despair of "this sort of thing is not possible in my school/with my sort of students/in this part of town/at this grade level/with this school administration/in this school board/in this subject area/with these parents/under these economic conditions," and so, on and on, we offer an old and pointed response of our late colleague, teacher and friend, Patricia Clifford: "if it actually exists, it *must* be possible."

So, then, here we are. We will do this.

"So, Here We Are"

The central motivating factor of this book is to elaborate beautiful classroom work that we have witnessed, over and over again, in every grade and every sort of school circumstance, over the past several years. To do this, we explore three interrelated roots of scholarly work that have a supportive and elaborative affinity to authentic and engaging classroom inquiry: ecological consciousness; Buddhist epistemologies; philosophies and practices; and interpretive inquiry or "hermeneutics." Although these three roots originate outside of and extend far beyond most educational literature, understanding them can be of immense practical importance to the conduct of rich, rigorous, practicable, sustainable and adventurous classroom work for students and teachers alike. They can help break the spell of educational discourse that has become moribund and stuck, and, worse, yet, dangerous.

Consider the words of David G. Smith (1999, pp. 135–6), one great teacher and refuge for us:

> "Education is suffering from narration-sickness," says Paulo Freire. It speaks out of a story which was once full of enthusiasm, but now shows itself incapable of a surprise ending. The nausea of narration-sickness comes from having heard enough, of hearing many variations on a theme but no new theme. A narrative which is sick may claim to speak for all, yet has no *aporia*, no possibility of meeting a stranger because the text is complete already. Such narratives may be passed

> as excellent by those who certify clarity and for whom ambiguity is a disease to be excoriated. But the literalism of such narratives (speeches, lectures, stories) inevitably produces a pedagogy which, while it passes as being "for the good of children," does not recognize the violence against children inherent in its own claim. Because without an acknowledgement and positive appreciation of the full polysemic possibility which can explode forth from within any occasion when adult and child genuinely meet together: a possibility which resides precisely in the difference of every child, every person, a difference about which one can presume nothing despite the massive research literature (e.g., about children) available to us, and despite the fact that our children come from us, are our flesh and blood. Without an appreciation of the radical mystery which confronts us in the face of every other person, our theorizing must inexorably become stuck, for then we are no longer available for that which comes to meet us from beyond ourselves, having determined in advance the conditions under which any new thing will be acceptable, and thereby foreclosing on the possibility of our own transformation. This radical difference of every child, every other person, renders our pedagogical narratives ambiguous but at the same time hopeful, because the immanent ambiguity held within them opens a space for genuine speaking, holding out the promise that something new can be said from out of the mists of the oracle of our own flesh.

These three disciplines of ecology, Buddhism and hermeneutics, each in their own way, knows something of this oracularity of the flesh, of the opening, in the concert between teacher and student, of "free spaces" (Gadamer 1986, p. 59) in intimate pedagogical acts of "responding and summoning" (Gadamer 1989, p. 458). Each allows insight into our frail human circumstance and offers ways to embrace and cultivate the wisdom of such frailty without balk or panic or denial.

Each of these three disciplines has cultivated, in its own way, a dual insight. First, each has carefully detailed ways to decode something of our contemporary lot in education: the wide-spread dominance of models of industrial assembly as befitting teaching and learning, the fragmentation of the living fields of knowledge, and the consequent acceleration and proliferation of demands on the lives and attention of teachers and students alike. Each speaks of distraction, of the leveling of experience and insight commonplace to everyday life.

In addition to such critiques, all three offer concrete, practicable alternatives to this fix we have inherited. They offer vivid ways of understanding issues of identity and diversity; they provide images of stillness and a slowing of time and attention linked to the pursuit of wisdom; they elaborate a sense of lineage, ancestry, or intergenerationality and the difficult comforts to be found in such elaborations; they detail ideas of living fields full of relations of dependent co-arising that can be explored, cared for, and understood; they confront head-on ideas of finitude and impermanence that are inevitably linked to the generative and ongoing character of knowledge and its pursuit. All three, therefore, offer a *pedagogy* that both decodes our current circumstances and provides what we suggest is vivid, rigorous, practical and scholarly alternatives to those circumstances.

We have witnessed in numerous classrooms how these matters actually work themselves out in the day-to-day life of classrooms dedicated to rich, engaging, deliberate work, and our book will draw upon and detail many real-world examples from our field work.

Hence our subtitle. Right in the midst of the often-debilitating contemporary circumstances of schools, we have witnessed the near-miraculous appearance of beautiful and engaging work across a wide array of classroom settings.

It has happened. Therefore, it must be possible.

"The Aviator"

The cover illustration is of a lovely painting by an Alberta artist, Connie Geerts (www.conniegeerts.com) entitled "The Aviator." In local lore, the Magpie is known as a smart thief. Connie's permission to use her work for our book summoned up the quiet demand of such a beautiful work—to remain with it, to look again and again, to become, slowly and surely, able to experience what is there and has been there all along. We're reminded of Hans-Georg Gadamer's (2007a, p. 131) recognition of the "joyous and frightening shock" that comes when the true message of a beautiful work starts to hit home: "You must change your life."

We have found something of this sense of aesthetic repose in the work we have done, over the past year, with a group of classroom teachers and administrators (some of whom co-wrote our Chapter 8), where examples of classroom work, or small and ordinary events, took on the character and repose needed to stop each of us in our tracks and tell us that the promises and presumptions that heretofore carried us through the days and hours of school would no longer do. This is the great and pleasurable work of hermeneutics, of ecology, of Buddhism, that we must, over and over again, face ourselves and our limited afflictions and, with each other's grace and aid, we can find a sustainable and joyous refuge in an every expanding sense of our "selves" as housed in a reality far beyond the meager panics we often are asked to take, in education, for "the real world."

For this wee little gift, we are most grateful, and the promise to write this book, now fulfilled, will hopefully provide a wee gift to those who read it and a wee sense of engaging readers, then, in an unspoken promise of their own with such reading.

After all, as a wise woman once insisted, don't tell me it's not possible. If it actually exists, it *must* be possible.

So, here we are.

Acknowledgments

We wish to acknowledge the permissions we have been given to reproduce previous published work:

CHAPTER 1: A Curriculum for Miracles by Jackie Seidel was previously published in Chambers, C.M., Hasebe-Ludt, E., Leggo, C., & Sinner, A. eds. *A Heart of Wisdom: Life Writing as Empathetic Inquiry* (2012). New York: Peter Lang, pp. 273–280.

CHAPTER 2: *Wabi Sabi* and the Pedagogical Countenance of Names by Jackie Seidel and David W. Jardine was previously published in N. Ng-A-Fook & J. Rottman, eds. *Reconsidering Canadian Curriculum Studies: Provoking Historical, Present, and Future Perspectives*. New York: Palgrave Macmillan, pp. 175–190.

CHAPTER 3: On the Spring-Squall Arrival of a Pine Siskin (*Carduelis Pinus*) by David W. Jardine was originally published in *Undivided: The Online Journal of Nonduality and Psychology*, *2*(1), www.undividedjournal.com.

CHAPTER 4: Reading the Stones by Jackie Seidel, was previously published in *The Canadian Journal of Environmental Education*, *5*, Spring 2000, pp. 178–185.

CHAPTER 5: Translating Water by David W. Jardine was previously published in the *Journal of Curriculum Theorizing*. *24*(1), pp. 11–19.

CHAPTER 6: Field Trip Curriculum by Jackie Seidel was previously published in *JCT: Journal of Curriculum Theorizing* (1999) *15*(4), pp. 155–156.

CHAPTER 8: We gratefully acknowledge Sue Goyette for permission to reprint "The True Name of Birds" and Don Domanski for permission to reprint "Disposing of a Broken Clock" in this chapter. We also acknowledge Kitty Lewis, the General Manager of Brick Books, for helping us secure these permissions.

CHAPTER 10: Inquiry in Black and White: An Appreciation by David W. Jardine was previously published in *One World in Dialogue*. *1*(1), pp. 33–36.

CHAPTER 11: The Paperwhite's Lesson Plan by Jackie Seidel was previously published in *Undivided: The Online Journal of Nonduality and Psychology*, *1*(3), http://undividedjournal.com.

CHAPTER 12: "The Memories of Childhood Have No Order and No End": Pedagogical Reflections on the Occasion of the Release, on October 9th 2009, of the Re-Mastered Version of the Beatles' *Sergeant Pepper's Lonely Hearts Club Band*© by David W. Jardine was originally published in the on-line *Journal of Applied Hermeneutics* at http://136.159.25.71/jah/index.php/jah/article/view/7/pdf.

CHAPTER 15: Some Thoughts on Teaching as Contemplative Practice by Jackie Seidel was previously published in *Teachers College Record, 108*(9), 2006, pp. 1901–1914.

1. A Curriculum for Miracles

JACKIE SEIDEL

> **Miracle**: From the Latin *mīrāculum*: object of wonder. *Mīrāculum* from *mīrārī*, to wonder at. From *mīrus*, wonderful. From *smeiros* [(s) mei–PIE–proto-Indo-European] "to smile, to be astonished." Also Sanskrit: *smerah*, "smiling." Also Old Church Slavic: *Smejo*—to laugh.

I began teaching in 1991 when I was 24 years old. I remember thinking that I might enjoy it enough to do it for a few years and save money for what I really wanted. Perhaps a Master's degree in English. Perhaps I'd be a writer.

Then, the miracles started to happen. I was overwhelmed by the intelligence and creativity of children and by what happened when we came together around shared ideas, readings, art, thoughts, our own life struggles and joys. From children, from being with them in schools in both friendship and fellowship, from walking the road of life together for one or sometimes two years, I learned to live life more graciously, more deeply, more slowly, more compassionately. With more breath. I learned to expect miracles and also to create space for miracles to happen. I learned that life itself is a miracle and that we are miracles, each of us.

Breath

We began the grade-three school year in mathematics by meditating on the meaning of the number one. We sat together, thought and talked for a long while, wrote many notes together. The conversation deepened. The children got excited. Their ideas grew and bounced from one to the other.

There is only one universe, said a child.

There is only one earth, responded another.

There is only one me.

And there is only one of each human person ever in the one universe.

One is unique! arrived a final comment. *Everything that is alive only happens one time ever!*

The children started to laugh. In this moment of thoughtful expansiveness a palpable ripple of delight flowed through our classroom as we experienced our mind's ability to have such thoughts.

Breath

It was recently genetically confirmed that the common maternal ancestor of all modern human beings lived approximately 200,000 years ago (Parry, 2010). One mitochondrial mother. One common human ancestor whose miraculous survival lives on in each of us. One unique wondrous human line of life.

The skin. The flesh. The earth. One well of water. One breath. Billions of years. We are the stuff of stars. Cosmic miraculous life. We are iron and calcium and oxygen. We are the rocks and the water and the wind. Breathing the breath of plants. What if such a miracle was the ground of our curriculum? Would we be inspired to perpetual reverence and awe for life, for one another, for our home?

As a teacher, I began to wonder how to hold on to such thoughts while terrorized by the stress of narrow curriculum goals or by those who deny the mysteries and miracles of life by seeking to quantify and know all things, or by those who seek to control teachers' minds and words with this or that method, or by those who seek to separate children into winners and losers, more and less, champions and failures, strong and weak. Deficits.

A Curriculum for Miracles is not a deficit curriculum. It is broad and wide and deep, holding the whole of life generously without crowding. Only in such a possibility do the deepest sorrow and suffering have a place, to be experienced alongside the greatest joy and transcendent, radiant peace. We know not one without the other. Exhale. Inhale. The outside is in and the inside is out. And the Western dualistic, individualistic mind dissolves into the miracle of life's flow.

Breath

Sorrow

This word appeared in the midst of reading a story together in our grade four classroom. In the midst of a grand conversation. Many of the children were learning English and we stopped to talk about what sorrow means. They gave examples. Jasmina raised her hand and spoke softly: *My mom has that. She cries all day.*

Something broke. The heart stopped for a moment. How could the tongue speak these names? Bosnia. Srebrenica. Genocide. Faraway places now so close, in the very skin and breath in our classroom. Again, it seemed so often, we were talking about war. It wasn't in the plans for that day. And then Samir was crying, too. His family also from Srebrenica. Their parents walked with such grief their bodies seem bowed to the ground. Other children nodded. The names of their stories sounded like Iraq, Afghanistan, Cambodia, Vietnam, Tibet, Somalia, Canada …

From such moments—many of them over many years—something was born and grew in me as a teacher. An invocation. A desire. An intention towards creating a curriculum of passion, creativity, happiness, spontaneity. Community. Love. Freedom. The movement of bodies and minds not stifled by hard desks and narrow ideas and rigid purposes. Laughter. Crying together. Memories and experiences cultivated, shared.

A sudden knowledge arrived one day, soaring into the classroom on some imperceptible updraft, freeing me from burdens I didn't know I carried. I knew with certainty (as much as there is any such thing) that all I could do as a teacher—nay, all I needed to do—was to prepare a joyous, creative day for children. Children who come as they are. Each of us a miracle of our ancestors' survival, whoever they were, wherever they lived. Here we are together today and *that is all.*

No matter the suffering they experienced in their day-to-day lives, empty cupboards, trauma and anguish, perhaps no one to love them enough, they could come to school each morning, each good new miraculous morning on this planet and have a wonder-filled day. A day in which their own unique potential and wholeness was fulfilled. A day of light and surprises and learning and enjoying one another. They could come to school and experience a Curriculum for Miracles. As a teacher, this I could do. And it is enough.

Breath

When I began this writing, I did not know this story now falling from my fingers.

Isaac, beloved nephew (son, brother, cousin, grandchild, great-grandchild), just turned twelve years old, loving, full of life, died early this summer. Our family had gathered for a rare time of togetherness to celebrate our grandmother's (mother, great-grandmother) birthday. A shocking, life-shattering, impossible accident interrupted and superseded the curriculum we had planned for ourselves.

This.

Sudden.

Absence.

The soft, soft body. The gentle, gentle breath.

Tender hearts break open in surprise. Rivers of tears. Salt of the earth. These cycles. Life.

Isaac was a miracle. A miracle in the middle of enjoying his creative, wondrous life. A miracle in the midst of reading books and working on his own projects. In the middle of making a *Fantastic Mr. Fox* claymation film. This was his important work in the world. Work that mattered.

A Curriculum for Miracles is a curriculum of the middle. Unfolding in the midst of things. In the midst of life and death. In the midst of joy and sorrow. In the midst of creation and destruction. It expands, always big enough, to hold the breath of wonder and the breath of anguish.

Many experiences as a teacher taught me that children's lives matter now and if there is any truth in the world, this is the one I learned to hold near to my heart: our fragile human lives are not in the future. A Curriculum for Miracles knows that life always unfolds from here and now, from this moment lived well with generosity and goodness. It does not waste children's time and it understands that each moment is important for its own time. This curriculum has no time to make children feel bad about themselves, about who they are, about their capacities as human beings. It only has time for having love, for being creative, for full and whole days of living life.

Breath

Four years ago I was working on a project with a class of grade-three students about the *United Nations Conventions on the Rights of the Child*. We studied the history and the wording of the Rights intimately. One day I asked them to write about the Right that meant the most to them. Their responses were overwhelmingly the same: You have the right to be alive. During this time, as part of my PhD dissertation-writing attempt to understand what it meant to be teaching in a time of ecological crisis, I was studying about the Chernobyl nuclear accident. I realized through this research that the future, in which all children have the right to be alive, in which all future children are born, will forever be a nuclear time. This was the most difficult knowledge I had ever carried.

As I watched the children in that class playing, learning and engaging in conversations, I recalled the words and stories of teachers and children who experienced and survived the Chernobyl accident. An unnamed child said:

> The sparrows disappeared from our town in the first year after the accident. They were lying around everywhere—in the yards, on the asphalt. They'd be raked up and taken away in the containers with the leaves. They didn't let people burn the leaves that year, because they were radioactive, so they buried the leaves. The sparrows came back two years later. We were so happy, we were calling to each other: "I saw a

> sparrow yesterday! They're back." The Maybugs also disappeared, and they haven't come back. Maybe they'll come back in a hundred years or a thousand. *That's what our teacher says. I won't see them.* (Alexievich, 2005, p. 218, emphasis added).

Lyudmila Dmitrievna Polenkaya, a teacher who survived Chernobyl and was evacuated from the Zone, described the experience for her as the opening of an abyss (Alexievich, 2005). Joanna Macy (2000) relates an experience where she met a school superintendent who carried a Geiger counter in his car so he could tell children where not to play. She met a school principal who had wall-papered his office with pictures of a forest, because they would not be allowed back into the nearby forest in his or even his grandchildren's lifetimes.

I thought of those teachers and children who faced this abyss. How could these teachers face their students after this? What did they believe they were now preparing children for? What teacher voice is possible in a post-Chernobyl world? I imagined being this teacher, telling children such a thing, that he or she would not see the Maybugs ever again. But then I thought of how this is happening today, and I am this teacher.

We studied animals together and the children asked about extinctions; they worried about polar bears and coyotes. They rescued spiders. We caught a wasp that had gotten into the school and put it in a container. We examined its hairy legs and body under the microscope, wondering at its miraculous eyes and wings. Some children were afraid of it. We had a conversation about keeping it to study—*it would die*—or about letting it go—*it would live. Let it live*, they said, *let it live. You have a right to be alive.* The children did not want life to die. Although they were still very young people, they were aware of the collapse of life systems that support the existence of living creatures on the planet. They were aware of the impermanence of life. They were aware of the miracle of life. And it was right there in our classroom. In the midst of it all.

Breath

A Curriculum for Miracles walks gently on this earth. It leaves footprints of love, compassion, forgiveness everywhere. It breathes the oxygen of life, photosynthetic miraculous gift. Its blood hums with iron from the rocks and oxygen from the air. It bathes and drinks from the one well from which all life flows.

A Curriculum for Miracles laughs at small ideas such as preparing children to be future workers in the global economy, at measuring children to one-size-fits-all. It knows such thoughts are ridiculous, hilarious, and leaves them behind. It knows that life is wondrous and takes up life itself as its topic. It knows that each life is but a brief flash. Unique. Alive now. Days, months, years. Immeasurable and impossible, yet here it is. And here we are together. There is no other place or time except this place and this time.

Breath

It is from those children who were always outside the boundaries of the discourses of "normal," competition and success that I learned the most. I grew weary of arguments for and against integration and differentiation. There is no argument. There is only yes and yes and yes. Yes to diversity. Yes to the fragile bodies and everyone being together. Yes to the infinite interconnected miracles of life on this planet.

We were sitting in a cluster together on the floor where our grade-four class gathered to share our ideas, thoughts and work. We were sharing stories we had written and Marie put up her hand and grinned. She rose up awkwardly and her friends carefully made room for her to stand. Her Down Syndrome health-related challenges were causing difficulty with her joints and balance. A thyroid imbalance left her listless, breathless, and feeling irritated with the world. She picked her way between the bodies of her friends sitting on the floor, swaying back and forth as she transferred weight from foot to foot, her upper body leaned forward with her hands outstretched in front holding the book she had made. With her back to me, I could see Marie's diaper above the waistband of her twisted and not quite pulled up sweatpants.

Sudden and unbidden tears prickled my eyes. This intimate, vulnerable moment. A ten-year-old's incontinence and 25 other ten-year-olds' acceptance of this as a possibility for being human.

Marie smiled at us and read her story aloud. It was a collection of shapes and some stick figures drawn with her favourite bright, scented markers. Mostly pink. Some letters she was learning were printed around the pages. Her voice was gruff and she spoke with few words, excitedly pointing to places on her pages. She shared her book, five whole pages, and then finished, stood proudly at attention. Waiting for a response. Expectant. Her friends burst into noisy and genuine applause and Marie's face flooded with overwhelming and radiant joy.

Throughout my days and years of teaching there have been too many of these miraculous wondrous moments to count. These moments offered lessons in being human, in finding humility in an institution that often has little, lessons in our deep and seemingly endless capacity for genuine love and care for one another, despite and across great differences in language, culture, religion, intellectual and physical capacity. Pedagogically I learned again and again that consciously creating a classroom that was a good and right place for

Marie, with enough space and time for a child like her to participate wholly and fully each day, all day, in everything we did together, was a classroom that was good for all children and also for their teacher.

Children like Marie brought me to a deep realization of the frailty of the human condition and the importance of radical diversity to all of our survival. They taught me to reject visions of education that did not include all children completely with rich purpose and full human experience. A Curriculum for Miracles is ecological, bursting forth from the understanding that the more diverse an environment is, the more creative and emergent the possibilities.

Breath

A Curriculum for Miracles breathes with a soft breath. And a fierce breath. Aware of the spark of life that flows everywhere at once and through all earth time. It knows that life is this fragile inhale and exhale that encircles the planet. It knows this breath is wondrous. It faces ecological crisis with courage and heart and knows that the sanctity and reality of death are always with us, yet also holds the spark of emergent life in its hands. It understands that schools as institutions habitually deny the imminent reality of death by casting life always into the future, and thus, it knows about time, and that breath is always now. And now again. A Curriculum for Miracles is one that breathes the present, holds itself intentionally close to the relational cycles of life (living and dying). It is awake to the profound, the mystical, the sensual. It responds to suffering of all kinds and cries out against injustice because it knows that each life is a life. Incarnations of breath flesh bone blood spirit.

A Curriculum for Miracles understands that life can be opened from this place called a classroom or school, or it can be closed. Life can be seen as wondrous or as dull. It can creatively overflow with joy, justice, peace and love, or it can serve the future, the literal, the non-miraculous. A Curriculum for Miracles knows that the latter are a path to the forever death called extinction, the end of miracles forever. Thus, a Curriculum for Miracles is a curriculum that knows that life itself is an Object of Wonder. Fragile. Unique. Interconnected. Just once.

A Curriculum for Miracles smiles and cries, dances and laughs. It is astonished. In the midst of it all. This is the curriculum that teaching has taught me.

2. Wabi Sabi *and the Pedagogical Countenance of Names*

JACKIE SEIDEL AND DAVID W. JARDINE

> *Wabi Sabi* is a way of seeing the world that is at the heart of Japanese culture. It finds beauty and harmony in what is simple, imperfect, natural, modest, and mysterious. It can be a little dark, but it is also warm and comfortable. It maybe be best understood as a feeling, rather than as an idea. (Reibstein & Young, 2008, [n.p.]).

I

These considerations began with our discovery of a new picture book entitled *Wabi Sabi* (2008), written by Mark Reibstein with stunning artwork ("collages" made from "a collection of time-worn human-made as well as natural materials" [cited from the inside back cover]) by Ed Young.

The book begins with a description of the meaning of the concept of *Wabi Sabi* (above), and a Zen proverb:

> *Kosho Hanaya Wo Danzu.*
>
> An old pine tree can teach you the sacred truths. (n.p.).

We turn the page and are ushered into the book's mysteries and exquisite beauty. Oriented as if a scroll, opening not left to right but top to bottom, four texts inhabit each page simultaneously: the narrative text of the story, an English haiku, a Japanese haiku (by Basho and Shiki, in Japanese Kanji, with English translations provided at the end of the story) and the collage illustrations.

Between these multiple texts, we enter the story of Wabi Sabi, a cat in Kyoto, Japan.

Near the beginning of the tale, strangers appear, "visitors from another country" (n.p.), who ask Wabi Sabi's master about the meaning of the cat's name. Her master responds, with an intake of breath, "That's hard to explain." The rest of the story unfolds from this moment, as Wabi Sabi experiences a profound unsettling of her identity: "It had never occurred to her before that *wabi sabi* was anything more than her name" (n.p.). Curious, she asks her animal friends about the meaning of her name. They, too, say that it is hard to explain. A passing bird tells her a monkey named Kosho can help her:

> A wise old monkey
> living among the pine trees
> knows *wabi sabi.* (n.p.).

Wabi Sabi sets off on a journey, through the dazzling, shining city and then to the woods of Mount Hiei where she falls asleep beneath an old pine tree. She awakens to the sounds of the monkey Kosho preparing tea nearby. To her query about her name, the monkey also responds "That's hard to explain," and invites Wabi Sabi to "Listen. Watch. Feel" (n.p.).

The monkey makes tea as if dancing, holding wooden and clay objects as if gold, and speaks to her in haiku. Wabi Sabi notices the designs of life in the woods and realizes that everything is "alive and dying" at once. Finally, Wabi Sabi comes upon seeing a reflection of herself in her tea comes whispers, "Now I understand." On her way home, "because she did not hurry," she finds Ginkakuji, the "Silver Temple" (n.p.). Inspired by the simple beauty, she composes three haiku before continuing on her journey. When she arrives home, Wabi Sabi curls up on her straw mat in the kitchen, warm, content, still smelling "the wind in her fur" and feeling her "journey's steps deep in her bones" (n.p.).

Her master is happy to see her: "Wabi Sabi!" she cries, "Where have you been?"

Wabi Sabi purrs, "That's hard to explain" (n.p.).

II

Seventeen years ago now, driving my [D.J.] ten-year-old son to school, deep winter days of overhead grayness.

Suddenly: "Dad, what would have happened if we had called trees 'weekends' and weekends 'trees'?"

Part of the pedagogical countenance of teaching is to love such moments, since they can open up, for teachers and students alike, a whole, wild, living territory in which much is to be taught and learned for all concerned. Even if this is not literally one's own child speaking, it is still one's kin, and this boy's words are still about words, about the very thing we teachers continuously teach with and about. There is so much here for a teacher to experience, to explore, to *enjoy*, in this happenstance question "what would have happened?" Here is a young boy innocently flirting at the slippage between names and things, foreshadowing all the Saussurrian arguments of post-modernism about how signs signify more signs, and always slip sideways away from the things we once so innocently thought they simply named.

But perhaps here is the rub, that something like this insightful Saussurian slippage is the source of the journey to find the calling of one's name. It is the impetus to find out what is called by *all those things* seemingly simply named in what was meant, from its name, to be a curriculum *guide*. Differently put, without this slippage, the name is just the name, pinned on the thing named, nailing it down, finishing it—something to be "covered." Without the slippage, the name loses something of its *calling* and becomes, like Wabi Sabi thought at the beginning of the tale ("it had never occurred to her" [n.p.]), nothing more than a name. Without the slippage, names don't need to be *heeded*. They don't beckon, and we don't need to *follow* that beckoning (what the name is calling us to, what it is calling is for, what our calling might be in being called by this name).

There is a story in this name.

Wabi Sabi discovered something hard to explain: to be so named might be a provocation, calling (in Latin *vocare*) something "forth" (Latin *pro-*), which, when elided, also means a "challenge" (Latin *provocare*). That is why she journeyed, because her name beckoned. Within such a slippage, there is teaching and learning to do—a "journey's steps deep inside [our] bones" (n.p.).

III

> In the earliest times the intimate unity of word and thing was so obvious that the true name was considered to be part of the bearer of the name. In Greek . . . *onoma* means . . . especially "proper name"—i.e., the name by which something is called. A name is what it is because it is what someone . . . answers to. (Gadamer, 1989, p. 405).

Wabi Sabi experiences this intimacy because her name becomes not just what she is called but that *by which* she is called, that which *calls her*. Her name is a lure, asking for her to venture through worlds of experience in order to be understood. Little wonder those she meets along the way—and she herself after she returns home—consistently answer the question of "What does Wabi Sabi mean?" with "That's hard to explain."

This is why Gadamerian hermeneutics moves away from the phenomenological fetish with the lived-experience (*Erlebnis*) that one "has," to a sense of experience as something you "go through," something undergone or suffered (Gadamer 1989, pp. 256–7). The German term *Erfahrung* ("experience") indicates something of a journey, a traversing (German: *Fahren*, "to travel"). Becoming experienced (German: *Erfahren*) becomes understood as the very sort of thing Wabi Sabi is undergoing. There is even a clue in the roots of the English term "experience":

> To become experienced means "to learn your way around," that is, to have ex-*peri*-ence (Gk. *ek-* means "out of," *peri-* means "around, as in the term "perimeter"—the "measure" [*metron*] of "around" [*peri-*]). (Jardine, Clifford & Friesen 2006, p. 207).

Names—even the ordinary word "experience," which everyone already understands and whose meaning "goes without saying"—have sometimes hidden or occluded motion and agency to them, something to show and teach, some path set out that needs to be *taken* in order to be *understood.*

All of the figures Wabi Sabi meets along her way, all of the places she visits, are moments that keep the sojourn going. None of them define the name and end the movement. The name remains alive, a living part of a living world.

Even when Wabi Sabi herself returns home and is asked, "Where have you been?" she answers, "That's hard to explain" (n.p.).

IV

> There is a dialectic to the word, which accords to every word an inner dimension of multiplication: every word breaks forth as if from a centre and is related to a whole, through which alone it is a word. Every word causes the whole of the language to which it belongs to resonate and the whole world-view that underlies it to appear. Thus, every word, as the event of a moment, carries with it the unsaid, to which it is related by responding and summoning. The occasionality of human speech is not a casual imperfection of its expressive power; it is, rather, the logical expression of the living virtuality of speech that brings a totality of meaning into play, without being able to express it totally. All human speaking is finite in such a way that there is laid up within it an infinity of meaning to be explicated and laid out. (Gadamer, 1989, p. 458).

No word, no name, arrives alone. Not only do words have motion and movement in them. Every word summons up and responds to the world within which it calls, sometimes myriad worlds, "interweaving and criss-crossing" (Wittgenstein 1968, p. 33) like the layered texts of this children's book. "Only in the multifariousness of such voices does it exist" (Gadamer 1989, p. 284). Sometimes incommensurate, sometimes antagonistic worlds can be called up all at once through the very mention of a word.

Every word calls up a world. To understand any word is a worldly venture. It is this "worlding" (Heidegger 1962) that provides the imaginal territory or terrain for the ensuing venture. Becoming experienced in the meaning of her name is possible only once her name, Wabi Sabi, is experienced as an opening into a world to be ventured out, into and through, a world which calls her forth (one etymological root, by the way, of the Latin *educare*, education). To understand Wabi Sabi, then, is to venture through the world of its calling and thus venture to become experienced in that world. It is a pedagogical calling, full of learning and teaching to come.

V

> . . . not "this is that" but this is a story about that, this is like that. (Clifford & Marcus 1986, p. 100).

This storied order of language denotes a deeply ecological order: these Pine Siskins at the birdfeeder live in a place in which Pine Grosbeaks are summoned up, and certain winters and rains. They do not properly exist as themselves and only then have relations of kin and kind and place:

> Seeds from pine cones are food for many songbirds such as the pine siskin, the pine grosbeak and the crossbills. Because the pine seeds are naturally shed through the winter months, the pine seeds are also food to many mammals in barren areas. (Beresford-Kroeger, 2003, p. 110).

It is not just that the Pine Siskins are "surrounded." There is a more difficult truth here that Wabi Sabi is discovering: the Pine Siskins, like her, "are" their surroundings. They are empty (Sanskrit: *Sunyata*) of a self-existence (Sanskrit: *Svabhava*) separate from their abode. This is why venturing through the abode[s] of one's name is required to understand "myself":

> One sees ones own self in all things, in living things, in hills and rivers, towns and hamlets, tiles and stones, and loves these things "as oneself." (Nishitani 1982, pp. 280–281).

This is "the self in its original countenance" (Nishitani 1982, p. 91). It is Wabi Sabi's "self" she is traversing, contours of her living, gatherings of her own countenance, steps inside her own earthly bones.

Thus, too, the name Pine Siskins "breaks forth," as Gadamer (1989, p. 458) puts it, "as if from a centre," just as these birds outside the window swoop and flit down in large flocks from the lodge-pole pine, just like these Old French (*loge*, "arbour" or "hut"), Latin (*laubia*) and German (*Laube*, "arbour") bloodlines flit around the edges of that name.

"It's hard to explain."

This is the tough insight that Wabi Sabi slowly happens upon: she, like the Siskins, like the lodge-pole, is the centre of a breaking-forth world of relations and, at the same time, "*the center is everywhere.*" Each and every thing becomes the center of all things and, in that sense, becomes an absolute center. This is the absolute uniqueness of things, their reality" (Nishitani 1982, p. 146). This is why, as the tale proceeds, Wabi Sabi can become so composed. She isn't lost or failing to be elsewhere. Home is everywhere, life is everywhere; the Pine Siskins matter as much (and as little) as the cones. They simply follow their ways in great concert with each other. This is the simplicity that Wabi Sabi seeks and means.

VI

This is also part of the sacred truth that the old pine tree teaches, and that it teaches us about teaching. It, like all things, "breaks forth". The Latin name for Lodgepole Pine is *Pinus contorta*. And this form of naming:

> presume[s] the great "Father of Taxonomy," Carolus Linnaeus (1707–1778), whose work, the enormous *Systema Naturae* (with its own bloodlines traceable back to the work of Aristotle before him), and his unfolding of the whole of creation into the grand typologies of Kingdoms and Genera and Species and Sub-Species and Families, that are still taught, in supplemented and modified forms, in our schools. From Linnaeus' binomial naming of these typologies in the One Universal Language of Latin, we have Latinate names that call back into a long colonial history which is not the one out of which this invitation had come to us. In a chilling parallel, and as was commonplace in his time, Carole Linne translated his own Swedish name to Carolus Linnaeus. Without such translation, one was not considered "a citizen of the world" (Reston, 2005, p. 127). (Friesen, Jardine & Gladstone 2010, p. 182).

There is another haiku by Matsuo Basho (1644–1694) about the pine tree that teaches about teaching:

> From the pine tree
> learn of the pine tree,
> And from the bamboo
> of the bamboo.

"Everything around us teaches impermanence" (Tsong-kha-pa [1357–1419] 2000, p. 151), and, to learn of *this* site of opening—the pine tree, the Siskins, the calling of the name Wabi Sabi—it is to this site one must go, and each of us must go there for ourselves and no one can spare us this venture of experience (Gadamer 1989, p. 356).

Wabi Sabi herself must go and no one can go for her.

VII

> The community is an order of memories preserved consciously in instructions, songs and stories, and both consciously and unconsciously in ways. A healthy culture holds preserving knowledge in place for a long time. That is, the essential wisdom accumulates in the community much as fertility builds in the soil. (Berry 1983, p. 73).

> Because she did not hurry, [Wabi Sabi] found a place called *Ginkakuji*, the "Silver Temple." (Reibstein & Young, 2008, n.p.).

There is/are another dimension(s) of the wor(l)d(s) that open(s) up for Wabi Sabi in the calling of her name: not only a lure to venture, not only surroundings which open, but an opening of time back into the ancientness that sounds in one's surroundings. This venture that Wabi Sabi takes is intimate to her, how she herself is called and yet, at the same time, it is intergenerational. She herself stands before those who have gone before, traces and tracks of paths through the city, through the woods, up to the old Temple full of the ghosts of those who have gone, and these are her teachers. Her venture is thus inherently pedagogical—in fact, early on, a bird hears Wabi Sabi's cries for help and says "there is someone who can help you" (n.p.). Not only does Wabi Sabi meet The Monkey, an ancient figure of a teacher, but this teacher teaches, not by telling her the meaning of her name but by making tea—a space and slow-time in which Wabi Sabi can compose herself and the experiences she has undergone.

This intergenerational character of the "breaking forth" of the name is again found in the language of Gadamer's hermeneutics. Becoming experienced (*Erfahrung*) is linked to a journey (*Fahren*) and these are both linked etymologically to those who have traveled here before us, ancestors (*Vorfahren*). This is another dimension of the wor(l)d(s) that open up in the calling of a name: not only a lure to venture, not only a surrounding which opens, but an opening of time back into the ancientness that sounds in one's surroundings. Paradoxically, the futurity that we experience when a name breaks open (that we have opening "ahead" of us multifarious ways that are both yet-to-be-ventured) is at once also experienced as full of the echoes of ancestral voices, ventures taken before, signs on paths, old texts, old stories, old figures. Basho Matsuo's (1644–1694) haiku appear as the pages turn, as do those of Masaoka Shiki (1867–1902), coiner of the term "haiku," just like we have left traces of H.G. Gadamer (1900–2002) here in this writing. All these names give an intergenerational surrounding to these tales. These, then, just like The Monkey in the Wabi Sabi story, are teachers who are invoked, those who have been on this sort of venture before us, teachers who are "kind," "kin." The ecological insight here is that we must also include those very Pine Siskins, who are themselves great teachers and who occasion an opening into ways that are themselves folds of "that anciently perceived likeness between all creatures and the earth of which they are made" (Berry 1983, p. 76).

Time cuts across this intergenerationalness in another way as well. Wabi Sabi meets The Monkey, an ancient figure of a teacher who teaches, not by telling her the meaning of her name but by making tea—an open, inviting space and slowing of time in which Wabi Sabi can, shall we say, collect herself. "Because she did not hurry, [Wabi Sabi] found a place called *Ginkakuji*, the 'Silver Temple.'" This lack of hurry allowed her to while away the time (Jardine, 2012e; see Chapter 7). In fact, the old temple entices Wabi Sabi not only to compose herself but also compose three haiku about yellow bamboo stalks, dark buildings and streams and leaves on a raked Zen garden. The writing of this children's book, the writing of this paper that is taking its own venture through the world opened up by Wabi Sabi—it seems like composition, composing oneself, is part of the journey-work that is being done to find the meaning of the name.

And thus, for us, is beckoned the gesture of writing. What might we tell our student-teachers about this book, if we ourselves have not taken the time to compose ourselves in the midst of the wor(l)d(s) it breaks open? We write, in some small way, in order to become experienced. But this sense of "being experienced" has the blush that Gadamer (1989, p. 355) adds:

> "Being experienced" does not consist in the fact that someone already knows everything and knows better than anyone else. Rather, the experienced person proves to be, on the contrary, someone who, because of the many experiences he has had and the knowledge he has drawn from them, is particularly well equipped to have new experiences and to learn from them. Experience has its proper fulfillment not in definitive knowledge but in the openness to experience that is made possible by experience itself.

VIII

> To begin a story, someone in some way must break a particular silence. (Wiebe & Johnson, 1998, p. 3).

> One day, visitors from another country asked Wabi Sabi's master what her name meant. It had never occurred to her before that *wabi sabi* was anything more than her name. (Reibstein & Young).

There is not simply an opening of time back into the ancientness that sounds in one's surroundings. There is as well, an opening in a future yet to be.

The arrival of strangers can interrupt all those hard-to-utter familiarities that define one's "home," one's "place," one's surroundings. As in this tale, the stranger can often ask the question of those very things that are beyond question and taken for granted by those who have become complacently at home in their surroundings. There are myriads of tales of strangers arriving and saying "goes without saying." The stranger often puts back into motion that which has atrophied, opens up what was a closed case, helps us remember something that was forgotten, awakens the life in something that seemed dead and forgotten. We even have in Genesis 17 an instance where the three strangers who visit Abraham are in fact heralds of the coming of new life, Sarah's impending pregnancy and the whole great line of ancestral surroundings that follows upon it. The strangers thus herald the intergenerationality of what once seemed barren. As go many old sayings, with the arrival of the new, the old becomes young again (just like Wabi Sabi's just-a-name seemed to wake up and become a new and fresh calling through a now-living world) and the fervency of the young—like Wabi Sabi's at-first-restless journey through this book—slowly comes to be held, comes to "find itself" and its calling—in the comfort of the world and its ways (which world, through the arrival of the new, as Hannah Arendt [1969, 193], can now be "set right anew"). The ancestral thus becomes *answerable to* the questions that the young pose to it and ancestry is called to remember, to reawaken, to open, to enliven anew its own wisdoms in such answerability.

The strangers thus prompt Wabi Sabi's journey by opening up the question of what Wabi Sabi means, a question which now gives her name, so to speak, a "future" that has yet to arrive. Something opens, like a "horizon of ... still undecided future possibilities" (Gadamer 1989, p. 112). This estrangement from simply "being at home" (the estrangement from deadening thinking that it is just a name) induces the task of becoming experienced. It induces, that is, a pedagogical sojourn, full of deeply Earthly, incarnate, sensual learning and teaching. It induces the need for new experience. And, as Heidegger (1962, p. 233) noted so well, the interruption by the strangers of the homely, "goes without saying" comfort of "it's just a name" gives way to what he called an experience of *unheimlichkeit*—literally "unhomelikeness." Instead of "home" being a given that is simply familiar, home has become opened up to the question of its countenance and continuance, what makes it alive and susceptible to a yet-to-be-determined future. The three who appeared to Abraham were not just strangers but *heralds*. They came with *news* of something *yet to come* (just like the experiences yet to arrive when Wabi Sabi first leaves her home). Home—like Abraham's bloodline—is thus held "open towards its future" (Gadamer 1989, p. 119) and Wabi Sabi becomes a calling which the strangers (the unhomelike ones, if you will) have helped "keep open for the future" (340) (notice how this places "diversity" and "difference" right at the heart of the health and well-being of "identity").

And this is not just "keeping myself open." This is a great pedagogical secret: there is no sense keeping myself open to a world which itself has no openings, no future, no possibilities, a world that doesn't call for my openness, my venture. Every student understands in their own way what it is like to be in a classroom where his or her presence cannot possibly make any difference, where everything is already decided, names ready to memorize, no memorable ventures to be taken.

The arrival of the openness of the young bespeaks the possibility of "keeping the world open" (Eliade 1968, p. 139) to the unforeseen possibilities of venturing.

IX

> Literalism is the enemy. Literalism is sickness. (Hillman, 1983, p. 3).

> The pine is first and foremost a tree of medicine. All over the global garden this knowledge is ancient. The pharmacy of the pine is as common in Turkey as it is in the Balkans, as it is to the Chinese, as it was to the ancient Picts of Scotland and now, as it is to us. (Beresford-Kroeger, 2003, p. 105).

> Throughout the ages the most popular use for pines was in the treatment of respiratory diseases. This included colds, coughs, laryngitis, chronic bronchitis, catarrh, sore throats, and asthma. This is because the pine exerts a dilatatory or opening action on the bronchi of the lungs. Since ancient times even a walk through a mature pine woods in summer was considered to be beneficial to one's general health. The fresh leaves exert a stimulant effect on breathing with the addition of mild anesthetic properties. There is possibly some mild narcotic function also in pines. In warm air the pine sweats a natural monomythyl and dimethylester of pinosylvin, which are both aerosols. (p. 109).

This is the pedagogical countenance of names. Just as the world helps Wabi Sabi understand the calling of her name, so too Wabi Sabi, in her world-opening ventures, saves the world of her name from being a "closed book"—she, like the strangers before us, asks the question which "breaks open the being" (Gadamer 1989, p. 360) of her name into its living world of living relations. Youthful, inexperienced, openness with no open world of ancient venture to take care of such exuberance afflicts the young with puerility. A "closed world" of already-foreclosed expertise is afflicted with senility, where Wabi Sabi's venture becomes little more than a bother. The pedagogical countenance of names is healthy and whole only in the properly pedagogical meeting of old and young, established and new, ancient and immediate. Without each other, each becomes ill, monstrous, distorted.

(Back in that car ride, with "weekends" and "trees," Wabi Sabi helps me understand why this speculative question of "what would have happened" is not an error requiring correction. It is a revealing of a difficult truth about words, an opening up of a way of words that is beautifully "hard to explain").

Healing is not a final state but attending after the possibility of continuance, able to face what comes, not imperviously finished with the venture but ready for it, not a stale expert but rather experienced in its ways.

From the Pine Tree learn of the pine tree, that it is a site where such haleness can be found, but it is not just healing *for us*:

> *Propolis* is collected from pines, among other species. The honeybee can be seen tearing at the resin with her mandibles. It takes her a quarter of an hour to load this *propolis*. When she arrives back at the hive, other worker bees help to unload. This is why spots of *propolis* are found around hive entrances. Bees mix *propolis* with wax and this becomes a fungicidal, antiseptic, and antibiotic wallpapering for the inside of the beehive. The fact that *propolis* is an old folk medicine in northern European forested areas is not surprising. (Beresford-Kroeger, 2003, p. 109).

X

> [It] compels us over and over, and the better one knows it, the *more* compelling it is. There comes a moment in which something is *there*, something one should not forget and cannot forget. This is not a matter of mastering an area of study. (Gadamer, 2007, p. 115).

The lure of the name *increases* as the journey proceeds, and there comes a moment in such ecological-educative movement "when something is *there*." We all understand this experience: the more I have come to know a work of art, a piece of music, a track of forest, a bird's call, the meaning of a name like Wabi Sabi, quadratic arcs, or the beauty of a beloved novel—the more I experience such things, the *more* compelling they become, the more they are experienced as "standing there," over and above my wanting and doing, there, in the midst of the world. The more I know about such worlds, the increasingly incommensurate is my own knowledge to the ways of that world. *It* gets better and bests my ability to outrun it with my knowing. *This* is what it means to become experienced. *This* is the pedagogical countenance of the name broken open.

Wabi Sabi? It's hard to explain.

3. *On the Spring-Squall Arrival of a Pine Siskin* (Carduelis pinus)

DAVID W. JARDINE

First, A Sorrowful Dedication

> *The whole leap depends on the slow pace at the beginning, like a long flat run before a broad jump. Anything that you want to move has to start where it is, in its stuckness. That involves erudition—probably too much erudition. One wants to get stuck in the history, the material, the knowledge, even relish it. Deliberately spending time in the old place. Then suddenly seeing through the old place.*
>
> —James Hillman (1998, p. 154), from *Inter views: Conversations with Laura Pozzo on Psychotherapy, Biography, Love, Soul, the Gods, Animals, Dreams, Imagination, Work, Cities, and the State of the Culture.*

> *A new idea is never only a wind-fall, an apple to be eaten. It takes hold of us as much as we take hold of it. The hunch that breaks in pulls one into an identification with it. We feel gifted, inspired, upset, because the message is also a messenger that makes demands, calling us to quit a present position and fly out.*
>
> —James Hillman (2005, p. 99), from *"Notes on Opportunism."*

James Hillman's work has formed, over the years, some of the imaginal spaciousness and cadences of my own writing. He's been one of those reliable refuges, because every time I read his work I know what's coming—a sometimes-hard-to-bear upset-summons into penumbral and often wondrous worlds. Breathtaking:

> *The word for perception or sensation in Greek was* aesthesis, *which means at root a breathing in or taking in of the world, the gasp, "aha," the "uh" of the breath in wonder, shock, amazement, and aesthetic response.*
>
> —James Hillman (2006, p. 36) from *"Anima Mundi: Returning the Soul to the World."*

I owe him some breath, this teacher I never met.

James Hillman died on Friday, October 28th, 2011 at age 85, and, in ways vital to my own work, I can attest without hesitation, and even at the distance of words, that he was much older and much younger than that.

So here's to flying out, again, in dedication.

I

> **Pine Siskin.** *Carduelis pinus.* A small, dark, *heavily* streaked finch with a deeply notched tail, sharply pointed bill. A *touch of yellow* in wings and at base of tail is not always evident. Most Siskins are detected by voice, flying over. (Peterson 1980, p. 272).

A single Pine Siskin is hunkered down in the corner of the feeder against these oh-so-ordinary Spring snow-squalls.

Foothills, west of Calgary, Alberta. March 24, 2012, first light.

Around 15 centimeters of "the wet stuff" overnight. Becoming a bit too damp for my aging desert bones.

Ecological awareness: this single Pine Siskin just happening to be there is an occasion for lingering ("residing, dwelling, lengthening" *OED*) over this arrival and trying to summon something beyond its ordinariness, and yet something right at the heart of that very ordinariness. This Siskin sits like a pre-dawn omen, a herald of a whole world of relations—this place, these snows, this time of year, this specific locale and terrain. And there, just coming into view under slowly gathering light, pressed-down snow hollows, hallowed markers, just like the tall-grass presses seen in summer, deft and familiar signs of a huddle of now-absent deer, last night in keep under the lowing of spruce branches, a bit nearer the house than usual, just to get out of the winds and the wet.

These spring squalls are, too, of a piece with the higher pitch of the sun these days, where the push of its heat is finally come back around, coupled, yes, with that strangely intense blue of the sky's Alberta arch.

This not a lifeless assembly gathered, here, in experience and memory around and in and through this Siskin. Each of these things is caught in relations of "responding and summoning" (Gadamer 1989, p. 458) to all of the other kin in this keep. Each beckons the other simply by being itself, because each already and irrevocably *is* all those only-*seeming* "others." We witness each other only because we are of the same earth-flesh. I see him there because I am visible; not self-contained or self-existent, and not even simply and only "open" but, phenomenologically speaking, *opened.* Thus a certain secret of hermeneutics:

> The lacuna, the weak place ... gives the opportunity. Perception of opportunities requires a sensitivity given through one's own wounds [the aesthesis of the "uh" denotes susceptibility, vulnerability]. Here, weakness provides the kind of hermetic, secret perception critical for adaptation to situations. The weak place serves to open us to what is in the air. We feel through our pores which way the wind blows. We turn with the wind; trimmers. (Hillman 2005, pp. 106–7).

This Pine Siskin is a gracious welcome to weakness.

II

But then the caution against the often-ridiculous parade of connectionism that comes from the first rush of insight into hermeneutics. Be wary of this giddy love affair. "*Poros* ... according to Plato's *Symposium* (203b-d) [Plato 2012b], is the father of Eros" (2005, p. 96). Openness fathers Desire, and the luring onrush that comes from the insight of hermeneutics can grab your snout, dim your wits, and run you ragged:

> Opportunities are not plain, clean gifts; they trail dark and chaotic attachments to their unknown backgrounds, luring us further. One insight leads to another; one invention suggests another variation—more and more seems to press through the hole, and more and more we find ourselves drawn out into a chaos of possibilities. (p. 99).

Thus a cautionary note regarding the alluring *spiel* of hermeneutic work: "One may become so engrossed in [possibilities and "connections"] that they outplay one, as it were, and prevail over one. The attraction [Eros] that the game exercises on the player lies in this risk" (Gadamer 1989, p. 106). This is the source of the great counsel regarding the cultivation of *practice* and the reason for many cautions regarding finding a teacher (Tsong-kha-pa 2000, pp. 69–92) who will not simply stimulate adventurousness—"the leap"—but also insist that there is value in spending time in the old place, value in returning there, nesting and nestling, that doubled helix of making a place whilst finding one's place, composing oneself in the fleshy composition of the world:

> The work ... is, as St. Bede suggested, a meditative, memorial activity, "a process of meditative composition or collocative reminiscence—'gathering,' *colligere*" (Carruthers 2003, p. 33). What is gathered in memory and experience must be worried over, worked, revisited, re-read, mumbled, "murmured" (Carruthers 2005, p. 164), so that when an occasion to write a paper like this one comes up, those matters can be called up from the recesses of memory, not as "raw data" but as inhabitants of places. In speaking of Quintilian, Carruthers (2005, pp. 297–8) insists that "composition is not an act of writing, it is rumination,

> cogitation, dictation, a listening, a dialogue, a 'gathering' (*collectio*) of voices from their several places in memory" (p. 3). "*Memoria* is most usefully thought of as a compositional art. The arts of memory are among the arts of thinking, especially involved with fostering the qualities we now revere as 'imagination' and 'creativity'" (Carruthers 2003, p. 9). Carruthers (p. 11) notes, too, that *invenio*, as the Latin root of inventiveness, is also the root of the term "inventory." Inventiveness and creativity are impossible if we have nothing to think *with*. (Jardine 2012d, p. 168).

III

Dependent co-arising (Sanskrit, *pratitya-samutpada*). Empty (Sanskrit, *sunya*) of self-existence (Sanskrit, *svabhava*).

Carduelis, from the Latin *Carduus*, meaning there might be thistles here and about, once Spring finally hits.

If it ever does.

And it always does.

And it will well beyond me and has well before. These old Lodge Pole Pines (expected companions to a Pine Siskin) are testaments to this.

Outplayed.

Mood and seasons in an old dance of tide pulls like the moon. And phases, lit and dark. Wane and wax. Attention and distraction. Dear me, this Siskin reminder. Where have I been, wasting this ebbing time, this opulent life of leisure and opportunity (Tsong-kha-pa 2000, pp. 117–128)?

IV

Any sense of self-existing "self" and "other" is a delusion aggravated by threat, real or imagined, snow-filled or moody, leading me to falsely grasp the very fleet of things, a foolish rush to stay these failings. The hermeneutic task is always one of how to write in such a way that those fleets are not betrayed and replaced with false confidences, but are, through composition, made available to lingering over, to being savored, experienced. The purpose of hermeneutic writing is not to give information, but to draw readers into "a hitherto concealed experience that transcends thinking from the position of subjectivity" (Gadamer 1989, p. 100; see Jardine 2012b). The purpose of hermeneutic writing is to cast both reader and writer into the open, the opportunity, of the matters under consideration, so that the life of those matters can be experienced and thought. There, in that open, both its and our fleeting co-failings can be brought close, held on the breath, trimmed.

An opulent and leisurely suffering. Leisure, Latin, *schola*, sensibly the root of "scholar," and ironically, perhaps even tragically, the root of "school."

And here is the moody whirlwind: that understandable-but-delusional grasp that sets up a false sense of inherent self-existence in the face of these vertiginous opportunities simply *increases* the susceptibility to feeling subsequently embattled, which thus increases the retraction into false promises.

And so on. A journal entry from February 22, 2012: "Now is not the time to try to grasp these ideas and threads and try to compose. Pulling at these tangled threads right now will only knot them tighter. Let them float by for now, noted."

Stuckness, for now but some days feels forever.

V

> Our specific human identities constructed through tribe, race or religion can never be ultimately secured, not only because they are always open onto the horizons of others but also, more important, because they are always already everywhere inhabited by the Other in the context of the fully real. (Smith 2006a, p. xxiv).

So, well, *think* and don't just wallow in the stuckness of the thickens.

My earthly body, curled up achy on this couch end, early morning, knows that this hunkered bird is not some isolated fragment, some separate and securable self-existence any more than I am, here, coffee and fire my hunkered refuge. It "is" only in and of the dependent co-arising of the places of its arrival, the fields of its turning, the worlds it inhabits and that, at the very same time, inhabit it. This ordinary and commonplace arrival of a Pine Siskin can, with practice, patience and forgiveness, be experienced for what it ecologically *is*: these worlds of relations, so to speak, *are* what it *is* its full "original countenance" (Nishitani 1982, p. 162).

That it appears *there*, tucked in that safest corner of the feeder against the sleet—that it appears as just this little bird—becomes world radiant in its very particularity.

Always, already, everywhere.

So, too, that very crack of firewood joins this gathering, as, too, upwelling pine smells pitched of last summer's splitting, still containing the warmth of the sun and the warmth of the memory of it.

That Pine Siskin sits as a constitutive feature of my own most intimate aesthetic inhale of breath at its sudden appearance.

It *is* that pitch of Pine, as am I.

We are each center and periphery to each other.

This paper is my warbling birdcall echoing back to it, in the scritching of a hand and ink over thick white paper, scratching and scrawling in the nest of words, making a nest in those nestles.

Quite unlike the furtive scurry-clacking of hard-shell beetles on a keyboard.

As per Roger Tory Peterson's (1980, p. 272) description of Pine Siskins, "most [writers] are detected by voice, flying over."

VI

> "It would not deserve the interest we take in it if it did not have something to teach us that we could not know by ourselves" (Gadamer 1989, p. xxxv). *It* has something to teach *us* because it speaks out of a cluster of occluded, perhaps unnoticed relations and dependencies we are already living within, that are *already at work* "before you know it," that is, before the deliberate deployment of methods aimed at controlling and managing its arrival. This sense of binding threads of obligation, commitment, implication, responsibility, and so on, is how the convivial world is experienced. We *belong to it* "beyond our wanting and doing," and aesthetic experiences make the living character and limits of such belonging experienceble, thinkable, utterable, and perhaps transformable. They make experiencable, thinkable, utterable and perhaps transformable who we understand this "we" and what we understand this "belonging" to be, and how we may have unwittingly limited or falsely presumed things about this "we" and this "belonging." (Jardine 2012b, p. 102).

And still, stop, *there it is*, "a being reposing in itself" (Gadamer 1977, p. 227), and, *in this very moment of reposing in itself*, it "breaks forth as if from a center and ... causes the whole ... to which it belongs to resonate and the whole world that underlies it to appear" (Gadamer 1989, p. 458). Its reposing radiance lights up "a world into which we are drawn" (Gadamer 1994, pp. 191–2).

Just this wee sentient being, just this little Pine Siskin, crouched, miraculous, *is* the "there" of such worlds, their event of "clearing" and "opening," as, too, is my attention to its arrival.

Like breath, outwards ("breaking forth *as if*") and then inwards (carrying all those relations back into its simple countenance).

Trimmed to the shapes of things and shaping things in such trimmers.

One gesture.

VII

"Everything around us teaches impermanence" (Tsong-kha-pa 2000, p. 151).

Only against the background of impermanence and emptiness is pedagogy possible.

"Everything around us teaches impermanence" (Tsong-kha-pa 2000, p. 151).

Otherwise, everything becomes presumed to be self-contained and nothing teaches, nothing learns.

"Everything around us teaches impermanence" (Tsong-kha-pa 2000, p. 151).

The great pedagogical lesson, then, is that *"the center is everywhere.* Each and every thing becomes the center of all things. This is the absolute uniqueness of things, their reality" (Nishitani 1982, p. 146).

"Everything around us teaches impermanence" (Tsong-kha-pa 2000, p. 151). ("*As if* from a center" [Gadamer 1989, p. 458]).

"The harmony of emptiness and dependent arising" (Tsong-kha-pa 2006) means that emptiness of self-existence is the same as exuberantly full of relations and dependencies.

"Everything around us teaches impermanence" (Tsong-kha-pa 2000, p. 151).

Emptiness is fullness.

"Everything around us teaches impermanence" (Tsong-kha-pa 2000, p. 151).

VIII

Another pedagogical lesson: "Cultivate love for those who have gathered to listen" (Tsong-kha-pa 2000, p. 64).

But this must issue from here: "Everything around us teaches impermanence" (Tsong-kha-pa 2000, p. 151). Otherwise it is a love based on delusion. Otherwise it is not love but rather *attachment.*

This love is *because of* impermanence. How could I ever love something that will last forever? Cultivate *this* love for that Siskin, for myself, for these words, for the heartache itself.

I *am* its being what it is and the inverse is also equally true: "one sees ones own self in all things, in living things, in hills and rivers, towns and hamlets, tiles and stones, and loves these things 'as oneself'" (Nishitani 1982, pp. 280–1). (But remember …).

Cultivate love for those who have gathered (but remember). Thus, too, my love affair with the harmonic tugs between writing and its topics, between writing and becoming more able to feel the fleeting fabrics of things: "I compose this in order to condition my own mind" (Tsong-kha-pa 2000, p. 111, citing Shantideva [see 2006]).

I love being amidst this "multifariousness of … voices" (Gadamer 1989, p. 284) and the comfort keeps of scholarship. Writing can be its own odd refuge, even when it must include the need to wait, to worry and to get dipped down into hallucinations of under-worlds and their grabs.

Writing is not a refuge *from* this: "Everything around us teaches impermanence" (Tsong-kha-pa 2000, p. 151).

It is a refuge *in* this.

I, too, this frail thing, have gathered to listen. To listen, I write, caught in the synesthetic cadences and countenances of little birds.

IX

I must cultivate a love of my sorry self and those sorrier afflictions and their seasonal returns:

> When we find ourselves in dangerous situations in which there are abundant stimulants for the afflictions, we should cultivate the antidotes to them with a proportionate intensity and we should stand up to them in a thousand ways. It is said that the best practitioners use as the path the very object that gives rise to the afflictions. Average practitioners apply the antidotes and hold their ground. Practitioners of a more basic capacity must abandon such objects and retreat. (Pelden 2010, pp. 253–4).

I must learn to love that I *am* best, average and basic, slip-slidy. This myriad of "ams," too, gathers around this Siskin.

To listen.

"Everything around us teaches impermanence" (Tsong-kha-pa 2000, p. 151).

I have birdcalls in me. And the skulls of our dogs killed by cougars are implanted near my spine, clipped on with hard cold metal spikes, forming part of my rib caging, sometime too closely enclosing labored pulls of breath.

"Everything around us teaches impermanence" (Tsong-kha-pa 2000, p. 151).

Opportunities are not plain, clean gifts. They include charnel reminders that hold at bay some of the silly romance of hermeneutics, of the task I face to cultivate love for those who gather to listen.

Everything around us teaches.

This, too, is aesthetic upwelling. I must learn to forgive myself for sometimes abandoning such objects and retreating.

Fucking cougars. They just won't listen. But that is how they listen.

Outplayed.

X

"Cogitation makes us expand, expansion stretches us out, and stretching makes us roomier" (Carruthers 2005, p. 199, citing St. Augustine).

"Slackness ... is an excessive withdrawal inward. You counteract it by stimulating the way you apprehend the object and by making the object of meditation extensive so as to expand your mind" (Tsong-kha-pa 2002, p. 63).

"Your mind becomes like a good field" (Tsong-kha-pa 2004, p. 38).

"Over and above our wanting and doing" (Gadamer 1989, p. xxviii), the in-breaking hunch of the world (Hillman 2005, p. 99), "uh" (Hillman 2006, p. 36):

> It captivates us just as the beautiful captivates us. It has asserted itself and captivated us before we can come to ourselves and be in a position to test the claim ... that it makes. In understanding we are drawn into an event of truth and arrive, as it were, too late. (Gadamer 1989 p. 490).

Always already everywhere.

And in the midst of this always-too-late-arrival, "something awakens our interest" (Gadamer 2001, p. 50).

Awakening, like this beautifully hopeless attempt to write in letters and words the sounds these Pine Siskins can make: "call, a loud *clee-ip*, or *chlee-ip*; also a light *tit-i-tit*; a buzzy *shreeeee*. Song suggests Goldfinch's: coarser, wheezy" (Peterson 1980, p. 272).

This is how to teach phonics, like the mind, itself a good field, full of the mongrel relations of English, full of old places and leaps.

Sometimes coarse and wheezy.

XI

> Some people are beginning to try to understand where they are, and what it would mean to live carefully and wisely, delicately in a place, in such a way that you can live there adequately and comfortably. Also, your children and grandchildren and generations a thousand years in the future would still be able to live there. That's living in terms of the whole fabric of living and life. (Snyder 1980, p. 86).

This place has *already* outlived me. This, too, is my "self" in its "original countenance," sitting squat in the middle of my own passing.

All things teach about the impermanence of *all things*.

There is an awful relief to be had in knowing that my self is no exception. That all life is suffering makes these passing sufferings more sufferable.

This, then, may be too dark a secret to utter: there is something of mortality at the heart of pedagogy.

And this utter locale imparts something of the "slender sadness" (Domanski 2002, p. 246) that comes from coming to know this secret:

> [It is] entering that state of being with a joy and wonder that comes from that very impermanency, from the absolute dispossession of everything we love and cherish. The wonder is that anything at all exists. The joy is that it does, even if it is as momentary as a human life. We can live this as a mode of attention, we can live within its movements, its cycles and treasure the phases, the round of it. (p. 246).

There is joy and wonder and love to be had in the round of this one Pine Siskin come heralding:

> The more [it] opens itself, the more luminous becomes the uniqueness of the fact that it is rather than is not. The more ... this comes into the open region, the more strange and solitary [it] becomes. There lies this offering, "that it be." (Heidegger 1977a, pp. 182–3).

"Just as the beautiful captivates us" (Gadamer 1989 p. 490), this Pine Siskin "attracts the longing of love to it. The beautiful disposes people in its favor immediately. It has its own radiance" (p. 481):

> We ought to be like elephants in the noontime sun in summer, when they are tormented by heat and thirst and catch sight of a cool lake. They throw themselves into the water with the greatest pleasure and without a moment's hesitation. In just the same way, for the sake of ourselves and others, we should give ourselves joyfully to the practice. (Pelden 2010, p. 255).

Postscript

> I have a friend who feels sometimes that the world is hostile to human life—he says it chills us and kills us. But how could we *be* were it not for this planet that provided our very shape? Two conditions—gravity and a livable temperature range—have given us fluids and flesh. The trees we climb and the ground we walk on have given us five fingers and toes. The "place" (from the root *plat*, broad, spreading, flat) gave us far-seeing eyes, the streams and breezes gave us versatile tongues and whorly ears. The land gave us a stride, and the lake a dive. The amazement gave us our kind of mind. We should be thankful for that, and take nature's stricter lessons with some grace. (Snyder 1990, p. 29).

Thankful, joyous, pleasures, grace. But, damn it, the snow is so thick and heavy, so full of disappointment. And that Siskin is so small and bereft. And my over-sentimentalized meditations, here, on the vagaries of its life and mine are both caught up, gripped right now with the tail ends of horrible dark affliction, seasonal, but not exactly a disorder—the Online Etymological Dictionary graces me with yet another grin, an term older yet less medically laden than "depression": Old English, *grevoushede*.

And a teacher I've never met has died and left me here, aging with a grievous head.

Again.

What is it that might be so wondrous, beautiful and joyous about such suffering?

Then this writing starts to end with perusing by chance a chapter from Gary Snyder's *The Practice of the Wild* called "Survival and sacrament" (1990a, pp. 175–185):

> Eating is a sacrament. The grace we say clears our hearts and guides the children and welcomes the guest, all at the same time. We look at eggs, apples, and stew. They are evidence of plentitude, excess, a great reproductive exuberance. And if we do eat meat it is their life, the bounce, the swish, of a great alert being with keen ears and lovely eyes, with foursquare feet and a huge beating heart that we eat, let us not deceive ourselves. We, too, will be offerings—we are all edible. (p. 184).

So, wee Siskin, this grace of writing is meant to clear our hearts, both of us, perishing as we are.

We are all edible.

Then noticing this, that Snyder has a header quote to this chapter (p. 175) from Dong-Shan (also transliterated as Tung-Shan [see Liang-Chieh 1986]), a 9th century Chan/Zen Master:

> One time when the Master was washing his bowls, he saw two birds contending over a frog. A monk who also saw this asked, "Why does it come to that?"
>
> The Master replied, "It's only for your benefit."

Carduelis pinus. Its huddled coldness *teaches.* However, as to the teachings of cougars, that lesson will have to wait.

That sadness is not yet slender.

4. *Reading the Stones*

JACKIE SEIDEL

This poem shares the story of a school camping trip to Dinosaur Provincial Park in southern Alberta. On the right hand side of the pages, in italics, are partial retellings of two ancient creation stories from Virginia Hamilton's (1988) *In the Beginning: Creation Stories from Around the World.* "Bursting from the Hen's Egg: Phan Ku the Creator" [pp. 21–3] is a Chinese story and "The Frost Giant: Imir the Creator" [pp. 68–71] is Icelandic. Some phrases have been directly cited from Virginia Hamilton's retellings. The children in my classroom who came on this camping trip, and where these stories were retold, were fascinated by the similarity between these two creation myths, particularly the notion of the god's death being the beginning of life, the god's body becoming the earth, the god's skull holding up the sky, and the images of continuous re-creation and rebirth.

. . . . . .

while walking
we find leaf skeletons
and wonder at the way
the life of the leaf has
dissolved become
earth

we cannot find even evidence
of the leaf in the soil but
our hands hold
this planet's breath

life structure fine bones so fragile remember
snows frost insects bite
sun   wind   rain   dark

fading light gives way to
gravity calling earth back to earth

a sudden genesis
something from nothing
exploding stars the fingers of gods
violently disrupt the dark

now then tomorrow
the gods walk the earth
and shout

crying lips pour forth surprised words

it is happening

watery burial spinning sun growing light shifting green
pulling moon tidal rush crushing sands moving rock plates collide
spewing fire volcanic ash growing life solid stone eating flesh
massive monsters drowning mud claiming life giving life

imagine

the first time we see
time's tattoo, the
fossilized evolutionary past

begin
breathe words
wind shifts
come life

seeds of life new yet already
history moving always
dying gods
gifting bodies to life

begin again

etched in our imagination
the mythic image
bone becoming stone
carrying life

in the beginning is no thing
it is something not yet born
it is called phan-ku

phan-ku emerges from the egg
carves the world with his chisel, holds up

the sky with his hands
the world not complete until he dies,

blood becomes water
filling rivers and seas,
flooding over the earth
bones become stones

skull holds up the dome of the sky

life emerges in the instant of death,

eternal resurrection to new life

our fascination with digging up
those bones far deeper than we know
is settled in souls intent on divining mysteries

are birds the living descendants of dinosaurs?
will the secrets be revealed in the excavation
of this earthly narrative?

begin

out of dust
bones
a rib flesh breath

beginnings rooted in soils

earth already witness to decay

we find another bone and everything
we thought
we knew
erodes
to new truth

begin again

in the yawning gap between north and south
between light and dark
that space where warm and cold meet,
the dripping water births the god imir
his licking tongue creates monsters from ice

then, murdered by his brothers
blood spills out
and out
and out

drowning monsters
filling the rivers, flowing to the seas

bones become stones
skull holds up the dome of the sky
decay breathes life

begin again

we walk the alberta badlands,
once swampy green fungus and monster filled,
and crush dinosaurs under our feet

we try to walk carefully but
there are too many bones and teeth,
fragments fallen from cliff faces, eroding into the rivers
joints from dinosaur knees and dinosaur hips waiting on rocks,
too big to be washed away
this summer
our hands hold them, gently

and when we lay down with our
noses to the hard-packed earth
we see millions of teeth
a sacred bone ground
everyone is quiet

wonder

this place,
once so alive and full
of breathing killing eating life

they fell by the waters

perfect conditions for a burial in sand and mud,
preserving connections through time,
bones becoming stones then waiting

the rain and mud now wash away the stone
reveal ancient lives and
move them away

if we come back tomorrow it will all be new
we cannot hold it here

more bones will wash down from the cliff
these too will pass into the waters

where, even now, more stone forms
below the crushing sands

the children call out as we are walking back
a different bone, a flat-disc, a sharp-edged tool
shines exposed in the harsh sunlight
who made this
shaping the bone
carving it with stone
scraping flesh from hide and flesh from bone

someone dropped this tool while walking through
walking home, stopping
perhaps to rest eat play love by this river

and now we leave this bone behind
in this hollow where we found it
more connected to this land than to us

it is not ours
but our fingers held it as fingers held it
long ago

it whispers a life story telling us
we are not the first to walk these steps
and crush these bones beneath our feet
under these wide, high, prairie-wind skies

bones and stones hold us together here
we feel so new

begin again,

and again,

ancient landscape of ocean beach
surprises us with shells above our heads
we see, walking past a cliff, imprints of life, millions of years ago
our fingers feel these shells and
we imagine sea-shore-waves
crashing
bringing them up
burying them
under the rolling, shifting sands and
we see shells today, now, in the middle of this dusty land
the wind almost sounds like ocean
as it blows through grass and hair

more, newer bones, shining white in the sun
near the path on the walk home

a deer was eating here
drinking from this now dry creek when water
ran in the spring
blinded by the clear bright light
a day like today
dangerous vision
a coyote bear cougar ate here too
leaving a sign of passing
only this carcass now
dragged through the grass by others
also passing

we wonder who shared this meal
or perhaps,
an old deer
lay down to rest here in this place
of bones
and did not wake in the morning

bones joining the stones
body giving life back
breath now wind
blowing over bones
whispers through dusty grass
now, tomorrow, yesterday, again

it is quiet around the fire later
our words left behind
back with those stones and bones

it is all too big for words
as we sit under the living trees
with the river running past
and the enormous star-lit sky
held up by the dome of our skulls

above
leaf skins brush together
skeletons rattling
wind-breath-whispering life

5. *Translating Water*

David W. Jardine

> Water is not only meant to reveal itself to the eye and the touch, but to speak and sing in seventeen different registers. Thus dream waters mumble and ebb and swell and roar and trickle and splash and stream and dally, and they wash you and can carry you away. They can rain from above and well up from the depths. (Illich 1992, pp. 145–6).

I

> All translation is interpretation.
>
> —Hans-Georg Gadamer (1900–2002)

> *Furuike ya kawazu tobikomu izu no oto*
> An old pond: a frog jumps in—the sound of water.
>
> —Matsuo Basho (1644–1694)

Matsuo Basho's (1644–94) old haiku is lovingly attended to in Hiroaki Sato's book *One Hundred Frogs: From Renga to Haiku to English* (1983). Along with placing the delicate art of the word and its translation in a tangled nest of wonderful historical, philosophical, cultural, spiritual, aesthetic, linguistic, and ecological contexts, Sato gathers together "one hundred frogs": dozens and dozens of suggested translations and takes on Basho's original work.

Already, the above citation (from Sato 1983, p. 149) from Basho betrays the fact that the original Japanese text, which is itself already a translation of a deeply meditative, lived-experience, is herewith *transliterated*. Already, the betrayals of words and their ways are deep and abiding. I will leave it to others to delve this depth. All I have for pedagogical experience, here, is the weird work of teaching young children whose tongues are other-wise that, in English, you have to start at the left-hand side of a word and move through it to the right in sounding it out. And this—"sounding it out"—needs to be done as a way of trying to work out a word you don't recognize "on sight," shall we say. As with, say, Japanese, "frog" is read by a reader "familiar" with this word, not as a string of soundable phonemes but as something recognized all at once, more like a picture seen in one glance and uttered in one sound, *frog*.

I must add, here, how fascinated a Grade 2 class was recently when we spoke together about the old, Early medieval quarrels about "silent reading" (very common in contemporary elementary school classrooms) and "reading out-loud." It is lost to memory that the very idea of silent reading entered into European consciousness around the 11th century (see Stock 1983; Carruthers, 2003, 2005; Carruthers & Ziolkowski 2002; Illich 1992, 1993; Illich & Sanders 1988), and has ancestries leading back to Augustine's idea of the "inner voice."

In almost all cases, texts were 'til then *voiced* when read. The idea of "silent reading" made no sense, since, without the voice's mutterings, without transport on the breath, without the spirit performing the text, the text remained dead and useless and meaningless. To read required that the text be inhabited by the breath of the one reading. This means becoming familiar with this habitat and its vestiges. We still recognise that some of us are better out-loud readers than others, and that to read silently something that one does not understand is one thing, but to try to read it aloud is truly strange and estranging. And all this is to say nothing yet about the differences, in elementary schools, between reading a story and telling a story.

Around the 11th century, two co-incident movements of thought occurred with a common consequence. Once written texts became more widespread, it became more possible to imagine the voices of the ancients housed in texts outside of myself. As such, as the ancients moved outwards into the world beyond my breath and voice, my sense of "myself" moved inwards. "Myself" became increasingly more singular, purged, less haunted by the ghostly voices of others. Knowledge became "out there" as I became "in here," and European philosophy was ripe for the moves of Cartesianism which fulfils the purging (the "de-worldling," disincarnating) of the self with a clear and distinct but empty "I am."

As "myself" becomes more intimately "interior," "silent reading" starts to make more sense. Moreover, as the voice moved inwards and reading aloud became less and less predominant, texts began to have to be punctuated, chaptered, headlined, paginated, and spaces began to appear between words. All this work had to become written which was once done by the out-loud reading voice. Thus began what Ivan Illich and Barry Sanders (1988) named "the alphabetization of the modern mind." Wonderfully, to name just these two for now, "silent reading" and "punctuation" in Grade Two are not simply language arts techniques to be mastered by children and bloodlessly inscribed as Curriculum Requirements. They are also lovely old stories about how things were once different about the voice, the breath, the sound of words, and the puncturing of calf-skins with inks.

II

> There is an old Italian saying: "*Traduttore, traditore* [to translate is to betray]." (Bethune 2002).

The last part of Basho's (already translated and transliterated) text is what is of interest here. A great ecological and meditative confluence: the sound of water and its translation into words.

Translation—it "betrays" something, hands something over one to another, it gives something away and takes something back. And, hermeneutically understood, this is precisely the roots, too, of those traditions which are not only handed over to us and in whose "handing down" (Gadamer 1989, p. 284) we are already inevitably involved, but to which, inversely, we have always already been handed over. Our very act of being human is already to be handed over, betrayed, visible and audible, presumed-upon, witnessed, not just witnessing, known, not just knowing. We don't begin as self-determining subjectivities but as already having been handed over to the ways of things (our language[s] and culture[s] and so on, all mixed and multifarious and, to the extent that we belong to them, often deathly silent and presumed). We are already betrayed by our belonging.

To *understand* this betrayal—to open them up to being other-wise—is the work of hermeneutics. To understand is to betray these betrayals, that is, to *interpret* them.

III

> What the expression expresses is not merely what is supposed to be expressed in it—what is meant by it—but primarily what is also expressed by the words without its being intended—i.e., what the expression, as it were, "betrays." In this wider sense the word "expression" refers to far more than linguistic expression; rather, it includes everything that we have to get behind, and that at the same time enables us to get behind it. Interpretation, therefore, does not refer to the sense intended, but to the sense that is hidden and has to be disclosed. The translator must preserve the character of his own language, the language into which he is translating, while still recognizing the value of the alien, even antagonistic character of [what is being translated]. (Gadamer 1989, p. 336).

Back to Hiroaki Sato's book, consider the intimate relation between the translationary betrayals of words and this living, Earthly presence, in Basho's words, of "the sound of water" (Hamill & Kaji 2000). Throughout Sato's collection, this last part of the *haiku* variously becomes voiced as "hark, water's music" (Bryan, in Sato 1983, p. 152), "the splash" (Miyamori, in Sato 1983, p. 152), "sleeping echoes awake" (Saito, in Sato 1983, p. 153), "plop!" (Blyth, in Sato 1983, p. 154), "the water's noise" (Fraser, in Sato 1983, p. 154), "WATERSPLASH" (Beilenson, in Sato 1983, p. 155), "a deep resonance" (Yuasa, in Sato 1983, p. 159), "a frog-leaps-in-splash ..." (O'Donnol, in Sato 1983, p. 159), "Kdang!" (Bond, in Sato 1983, p. 160), "water-note" (Maeda, in Sato 1983, p. 160), "with splash-splosh" (Ikeda, in Sato 1983, p. 161).

There are dozens more that are offered. From Allen Ginsberg, from whose work we've come to expect a complex relation to such matters of poetry and sound and the Beat-East, we get (I'm tempted to say "of course"), "kerplunk" (Ginsberg, in Sato 1983, p. 164):

> **Ker**- The first element in numerous onomatopoeic or echoic formations intended to imitate the sound or the effect of the fall of some heavy body, as *kerchunk, -flop, -plunk, -slam, -slap, -splash, -souse, -swash, -swosh, -thump, -whop, etc.* **U.S. vulgar**– 1903, in *Outing* XLIII. 83/1 "The sound made by the water when the frog dives, we used to express when we were boys, by the word '*kerplunk.*'" (Online Etymological Dictionary)

One could also consider trickle, slosh, plash, popple, ripple, burble, purl, gurgle, swash and murmur: various English soundings of "the play of water" (Gadamer 1989, p. 332).

Surely, in this *haiku,* these final words also *mean something* (breaking the surface, interrupting the stillness, opening the depths, or perhaps the sudden crack of sound that announces enlightenment after the stillness of the pond is broken ["breaking open the *being* of the object" as Hans-Georg Gadamer (1989, p. 360) describes interpretation's betrayals]). But still there is (to coin a phrase) a "plop" in which the ear abides.

Remember with those boys in 1903 the great turbulent thunks of a flat stone tossed high and entering water perfectly straight at a great Pythagorean right-angle?

An air-capturing ga-goomp like a bullfroggy throating?

And how my using these words this way betrays a different world than Bryan's (Sato 1983, p. 152) already cited neo-Victorian "Hark." What is betrayed in translating the sound of water into words is not just the sound of water but the sounding of the words themselves, portrayed, in this citation, with the sound-word "resonance":

> Every word breaks forth as if from a center and is related to a whole, through which alone it is a word. Every word causes the whole of the language to which it belongs to resonate. (Gadamer 1989, p. 458).

IV

> The French have a saying: "*Traduire, c'est trahir*—to translate is to betray." (Double Birdie 2005).

> The sound of water implies ... the eye and the ear of a recluse attentive to the minute changes in nature and suggests a large meditative loneliness, sometimes referred to as *sabi*: the sound of the water paradoxically deepens the sense of surrounding quiet. (Shirane 1996, p. 51)

> **NUMBER**: 38848
>
> **QUOTATION**: "Meditation and water are wedded forever."
>
> **ATTRIBUTION:** Herman Melville (1819–1891), U.S. author. Moby-Dick (1851), ch. 1, *The Writings of Herman Melville*, vol. 6, eds. Harrison Hayford, Hershel Parker, and G. Thomas Tanselle (1988). (http://www.bartleby.com/66/48/38848.html).

Out for a walk by the Elbow River and the various creeks that trickle into it. Remembering another of Basho's great invocations, this time to the pine tree:

> From the pine tree
> learn of the pine tree,
> And from the bamboo
> of the bamboo. (see the spectacular http://www.ahapoetry.com/haiku.htm)

And so too it is with the sound of water. Basho's haiku-invocation is a chance to remember that Earth-places can be great teachers, that there is learning to be had in the terrible presence of things and their ways. There is perishing here in this walking meditation that hears, a great sense of *passing*:

> If there are such things as natural symbols, then sounds are surely the natural symbol of transience and the lostness of past time. They are essentially evanescent, an exact correlative of wistfulness and poignant regret, not to mention sentimentality. They seem to be nature's way of mourning. (Ree 2000, p. 23–4).

Again, a great Medieval debate. To read out loud is to interpret, because casting written texts up into the voice is an act of incarnating, enspiriting and bringing to life what it is saying to us ("awakened into spoken language," as Gadamer

[1989, p. 394] puts it). Reading a text out loud means that I (and not just the text's "author") am *saying* these words. Something happens when we read something aloud. The voice is asked to experience the truth of the words in uttering them, and that truth is carried on a voice full of perishing and mourning and lostness, even when, perhaps especially when the words sounded speak to a truth that will outlast the breath of that frail voice itself. The voice and its sounds "passes by" like texts do not. The voice and its breathing pass away into silence. The airs stop moving, even while the written text remains, now the corpse of the vanished breath.

Where water sounds, water breathes. Hearing the sound of water is hearing the breathing (aeration) of water:

> a frog-pond ploomp!
> makes it breathe. (Flygare, in Sato 1983, p. 167).

To hear the breathing of water is to be one who breathes:

> Seeing the frailty of your life through seeing the breath is the meditation on the recollection of death. Just realizing this fact–that if the breath goes in but does not go out again, or goes out but does not come in again, your life is over—is enough to change the mind. It will startle you into being aware. (Chah 2001, p. 44).

Soundwalk near the river's edge with a difficult task in mind, and as a form of meditative obedience, to hear or to heed such perishing—the "emptying out" of things beyond their feigned and timid self-containment, out into all their relations. Caught in the sounding bristles of water's tricks over rocks, listening to the auditory spaciousness of the place and how the soundplays of water play out a huge, sensuous, multifarious voicing. This sound of riverwater sounds the distance of that rock face on the far side and later, as the face slopes downwards and away, the sound sounds this movement of rock (which is at once a movement of an animal body past such rock movements. This is *one thing*, not two) and can be heard to belong properly to it. Listening to its shifts and flutters as we walk—shifts in how these sounds are spread out territorially, marked around this animal-body and its bi-aurality: lefts, rights, distances, closeness, aheads, behinds, echoes off of steep rockshorelines. The sound is framed by distant read squirrel chits and chats whose scold is not about us. There's something up over there—bear? Other hikers? A hawk perhaps? Riversound. Woodswater. Pinepitched. Mourning. That heart-breaking sound of an unseen Red-Tailed Hawk overhead downstream has already disappeared without a trace.

> (An asthmatic squeal, *keeer-r-r* [slurring downwards]) (Peterson 1980, p. 154).

Sharpness of small fast bitty-trickles make breath rush a bit in a new wash of sounds. Listening to the breadth of this river's sounding is listening to a great three-dimensional space that surrounds this body. *This* body, here, now, full of aches, and now, in writing, remembered specifically in attempts to compose this waterwalk in the composures of writing.

But this is not quite yet a good betrayal of the sound of water. Consider: that these specific soundings, to be *just thus*, require *just these* rocks placed *just so*. These exact sounds require *exactly this* relation between gravels and shallows and high-pitched trickles. This is the sound, not just of these gravels, but of their having arrived here, with all the flooded stormwearing meticulousness that requires, with gravities and icecold rockbreakings and the shatters of falling cliffpieces. All of this is what this sound *is*.

This is its betrayal.

Further along, *these* exact sounds require *exactly this* placement of large rocks that can capture drumskins of air into deep, hollow-sounding adumbrations. And all of this requires all the ages of glaciers and plate shifts and spring run-offs and water-wearings and those cold ice Alberta winters and bear scramblings that, over a vast and patient time, placed just *that* pebble *there*. Consider: what am I *hearing* in the sound of water? It is an abstraction to think of sound waves and auditory canals alone (recall, however, that auditory canals are themselves partially swirls of water's sound-bearings). It is equally abstract to think that I am not hearing the ancestral ecological voices of this whole place, echoing just here, just now, in all its frail and passing particularity.

Shirane's *sabi*: a small, delicate water-sound like the lap at the water's edge can only sound in an acre of quiet. The smaller the sound, the larger the quiet must be. The smaller the sound, the larger the quiet becomes. *Hearing* the rock ancestries of this small sound "deepens the sense of surrounding quiet." It is the rocks-having-happened-to-fall-here sounding:

> Stillness and activity are actually the same thing. This short poem demonstrates that if there were no stillness to the old pond, there would be no sound as a frog jumps into the water. The activity exists as the same moment as the non-activity, they are the same thing. (http://openpoetry.com/BashoMatsuo).

It is ages sounding, just right here. It is all this handing-down that is betrayed. This trickle of water greened from the mineral-spring richness, *this* trickle betrays all things.

This is called "ecological awareness."

V

> License is not precisely betrayal, but another kind of faithfulness. But from the point of view of fidelity, understood as being bound to the literal text, it may well be that that other order of faithfulness, the one associated with freedom and license, can only be read as betrayal. (Butler 2004, p. 82).

Every translation of the sound of water into words, as with every translation of the words for the sound of water into another tongue—every translation is a betrayal, an interpretation which breaks open the being of the object and makes it vulnerable to the otherwise ear and tongue and imagination.

However, as Judith Butler hints, this need not be understood only negatively, as if human language somehow necessarily distorts or despoils the immediacies of ecological experience (standing here, body-facing, sounding water all around). Every translation of the sound of water into written words is also capable of *betraying something* of the sound of water, that is, revealing something, making something *show* about the sound of water, as much as leaving the sound of water still unsaid. It is easy to imagine that the betrayal of words means simply that words fail and do badly by things (I've often wondered if this is what David Abram [1996] is suggesting, paradoxically given how beautiful a writer he is). I'm suggesting that in their very failure to capture and claim and the thing altogether (in their failure to make the thing face this way and no other), they succeed in presenting the sound of water that, even though it is uttered in words, remains experienced, in these words, as *there*, "beyond my wanting and doing" (Gadamer 1989, p. xxviii), "standing there" (Heidegger's *Da*) reposing beyond the words themselves:

> The existing thing does not simply offer us a recognizable and familiar surface contour; it also has an inner depth of self-sufficiency that Heidegger calls "standing-in-itself." The complete unhiddenness of all beings, their total objectification (by means of a representation that conceives things in their perfect state [fully given, fully present, fully presented, fully written or spoken, finished]) would negate this standing-in-itself of beings and lead to a total leveling of them. A complete objectification of this kind would no longer represent beings that stand in their own being. Rather, it would represent nothing more than our opportunity for using beings, and what would be manifest would be the will that seizes upon and dominates things. In [hermeneutic experience] we experience an absolute opposition to this will-to-control, not in the sense of a rigid resistance to the presumption of our will, which is bent on utilizing things, but in the sense of the superior and intrusive power of a being reposing in itself. (Gadamer 1977, pp. 226–7).

Cultivating this experience of the repose of things is what hermeneutics calls "the art of writing" (Gadamer 1989, p. 390). In regard to the arguments regarding silent reading and reading out loud and speaking, Gadamer suggests that some writing that is well wrought "reads itself," (p. 390), writing which "draws readers into the course of thought," (p. 390), its "productive movement" (p. 390) in which the art of writing appears artless and disappears in favour of the appearance of the thing itself.

Moreover, every translation of the sound of the words for the sound of water into another tongue *betrays something* of the life of the words translated (in *both* tongues—"Hark," "kerplunk") and of the sound of water (it *shows* something about words and tongues and water's sounds and how each tongue sings such sounds out loud). Words are not *representations of things.* Words are not *stand-ins.* Words that bespeak the sound of water are meant to make it present, to show it off, to lead us to it and offer us up to its ways, not to stand in front of it and block our way. They are not substitutes but rather heralds of the arrival of the thing.

Words are another kind of faithfulness, a presentation of water's sound, a voice of the voicing thing.

In words, the thing appears. It is not just *referred to.*

VI

> Jane Reichhold, [see http:// www.ahapoetry.com/haiku.htm], who fires off missives on *haiku* over the Internet, claims that Basho's final phrase can be literally "water of sound" in Japanese: "The water of sound. Sound as water. Sound moving as water does. Sound rippling outward as water does when disturbed." She's suggesting that Basho's concluding phrase is an actual visual image—in water—of how sound moves. But I noticed—amazingly could feel—that the final Japanese words make an auditory equivalent of how waves of sound/water circle outward from their source fading as they go: *mizu-no-oto.* Do you hear those opening ripples in the repeated o's and in their duration? O's don't cut off like the p's in "plop!" They fade away. Like a *haiku* voiced out. (Bakken 2003).

VII

> There is an old Italian saying that equates the translator's craft with treason: *"Traduttore, traditore"*. A French version, laced with misogyny, suggests that a translation's fidelity to the original is inversely proportional to its aesthetic value: "*Les traductions sont comme les femmes, ou belles ou fideles* ["Translations are like women, either beautiful or faithful/true." (Gurria-Quintana 2006).
>
> Even if translation is treason, it is a necessary form of treachery on which readers depend. (Gurria-Quintana 2006).

Swallowing
Life rafts of pain pills—
with sips of chills

Basho's frog
floating in a jar of rain—

Hortensia Anderson (2007)

Final Pedagogical Reflection

> once upon a time ther was a rain drop and it gope on a bird then the sun trd into a watrvapr the radrop fad his bovrsrs and trnd into a fofe white cloud and then it trnd in too a havie plak kloub and then it trd in bake to the sam radrop and gropt on the sam bird.
>
> Name Eric

Below is a translation which opens this water-text out into a field of "conventionality" and allows its meaning to become visible and audible while, at the same time, betraying its frailty, this young boy and his ear for the soundings of words and their meaning—six years old at the time, writing at the computer—now slipped away into adulthood:

Once upon a time there was a raindrop.
And it dropped on a bird.
The sun turned into a water vapour.
The raindrop found his brothers
And turned into a fluffy white cloud.
And then it turned into a heavy black cloud.
And then it turned back into the same raindrop
And dropped on to the same bird.

Eric Jardine

I cite it here to be read out loud. It is my chance to mourn anew the sound of water and its passing.

6. *Field Trip Curriculum*

Jackie Seidel

Field Trip Curriculum I

the boys baptize their bodies
with mud down by the river, their white
teeth and eyes flashing with laughter and hysteria
and as they stomp and slosh and
unsettle the earth the softening mud
rises, squishing through
their toes, past
their ankles, half-way
to their knees and with
mudballs flinging and slinging and clinging in their hair
they wave mud-gloved hands   hey! look at us!
then they float silently, faces to the warming sun, drifting
with the current that runs through the prairie and on down

all the way to
the mississippi
clean and free

Field Trip Curriculum II

starlight ruminations and
windwhipped sandstone
silty river swimming

snakes curl on sunny rocks blue skies over
eroding cliffs' shadows

we walk together under petroglyph carvings
reading time's traces
etched in stone the
dog days and
then horses and
a battle and
a snake curled in the summer sun

past an arrowhead

shell fossils and sandy sedimentary pressure crushing ancient
beach becoming
rock

life's layers exposed

the northwest mounted police signed their names here

and then, a woman, upside down, child born feet first midwife
between her legs   what is time   under these blueblue skies
a story of death   life on its head
                     who wrote it?
someone cries for her

Field Trip Curriculum III

After marshmallows smoky fire and hot chocolate too much sun and journal writing we walk in the nightbright moonlight seeking solitary space one by one the places call us off the path carcass of a deer ribsfleshbloodboneflies surprises us around the corner and we sit on the edge of cliffs edge of the earth feet dangling dangerously quiet and alone but not really the night is loud and full and comes to meet us here we abandon ourselves to stars infinity strands us on this ancient shore laying on fossil layers of beach crushed sand we are swallowed and time bound where does it end and the prairie grass exhales sweet air caresses us good night.

7. *Story-Time Lessons From a Dog Named* Fideles

DAVID W. JARDINE

> [Storytelling] is a fluid tradition that is as migratory as a winter bird, feeding as it goes from place to place and leaving something of itself behind. Those of us with gardens can attest to the hardiness of "volunteers" that spring up from seeds that have been carried in a bird's body over countless miles.
>
> —Jane Yolen (1988, p. 3), from *Favorite Folktales from Around the World*

A Brief Story to Begin With

The King fidgeted. His faithful dog Fideles would always warn him if danger was near, if betrayal was sniffed.

The sun slowly warmed its way up between the turrets. The castle stood firm, flooded with new morning light that crackled the dew across the nearby meadows and the surrounds of fields already full of furtive work, busy hands, and work-songs lilting between birdcalls and swoops of wet, green air.

The Queen, finally, slept.

But not the King. His trust had slumped down like old and stagnant moat water, thick, scummy, without current, then suddenly jolting him awake him before sunrise with dreams of drowning in murk.

Was it her doing?

Was it the boy Prince?

One of the twin Princess sisters?

Or both! Duplicity! And what of that smoke spiral off in the distance beyond the ramparting arch of hills to the South?

"Pray, tell, Fideles" whispered the King into a long warm dog-ear. "Who or what is behind this lurking feeling that something is about to go horribly wrong?"

The Queen's cat murmured in a hidden curtain fold, tail switching, a cold ear turned to pointed listening, this time in secret. Old Thomas, much gray hair bristling now 'midst the once all black, would soon tell the Queen of the King's lamentations. He would have to wait until she awoke that afternoon. Waking her too early was always a risky prospect.

It is this telling and the troubles that this telling foretells that are the story of this tale. Gather. Listen.

But shush! Be still! The door! Fideles gives out a low growly gurgling meant to warn amid bristling hackles.

The King turns, the back of his neck, too, chilled and bristly. He had hoped this time would never come.

But here it is and here we are.

Sit! Listen! It has come ...

A Second Story to Continue the Tale

I've told this story and stories like it to the youngest of children countless times. There is something palpable about such gatherings, if well conducted, that supersedes teachers' or parents' fears of literacy skills and all those other easily induced panics that have ravaged too much of the great and ancient tasks of learning our way around language with children and in their midst. It is not that skills with letters and sounds and grammars are of no concern. It is that they have no legitimate claim to being "basic" or "fundamental," no claim to being first and foremost, either chronologically or otherwise. It is only fear and inexperience that have allowed them to bully their way to the front of the line, only an amnesia about stories and their charms that concede to this falsehood.

"Something awakens our interest—*that* is really what comes first!" (Gadamer 2001, p. 50, emphasis added). Something awakens our being in the middle (*inter*) of things (*esse*), and we find that there is a story already underway, one in which we are already moving and living:

> Told and retold or read and reread, the story exists neither in the mouth nor on the page, neither in the ear nor the eye. It is created *between*. No two listeners hear exactly the same tale. Each brings something of himself to the story, and the story is then re-created between the teller and the listener, between the writer and the reader. (Yolen 1988, p. 4).

Thus storytelling is linked up to the great arts of interpretation. It is the active and creative of weaving a text, a fabric (Latin *textus*) of experience and venture, between those gathered (tellers and listeners), the topographies of the story itself, and those whose tale the story tells. "*The true locus of hermeneutics is this in-between*" (Gadamer 1989, p. 295). Stories are always, even in some small ways, the stories of those who have gathered in the light (one could say "radiance") of the story being told.

Over the course of a story's time, listeners, and the storyteller, too, slowly start to "recognize themselves in the mess of the world" (Hillman 1983, p. 49, from a book tellingly named *Healing Fiction*) that is being unfurled. Like in the making of a fabric, each person gathered experiences their own bias in the bias fabric itself—and this seeming paradox ("the bias of the fabric"/"the bias of the listener") is part of the story's hold and why we are held *together*. The fabric pulls each of us differently as we each pull at it from here and there, and it also holds us together at the same time. This "at the same time" is part of the odd temporality of stories and their telling.

With lights lowed, we all lean inwards in this circle of storytelling, and its arc goes far beyond just those gathered here. This wider feel of fabric is also a nebulous part of the story told:

> Next to the hearth, by the bedside, on the back porch, round the cracker barrel, in the lap. Mouth to ear, mouth to ear, over and over and over again, grandmother and grandfather, uncle and aunt, mother and father, nanny and nurse were in turn listener and teller. (Yolen 1988, p. 12).

Migratory arcs. This temporal fabric is *recurrent* and *intergenerational*—an odd experience of going somewhere new and returning to somewhere old at the same time, and coming to (re)inhabit a place already inhabited.

> These tales could be a short as the English ghost story reported by the venerable English collector Katharine Briggs: *He woke up frightened and reached for the matches, and the matches were put into his hand.* (Yolen 1988, p. 2).

Or they could be as quickening as Jane Yolen's own brief tale about the comforts of stories themselves:

> "Story," the Old Man Said, looking beyond the cave to the dragon's tracks. "Story is our wall against the dark." He told the tale: the landing, the first death, the second. They heard the rush of wind, the terrible voice, a scream, then another. Beyond the wall, the dragon waited but could not get in. (Yolen 1998, from the back cover).

Comfort: common strength. This is part of the work of teaching, to cast a tale where no one is damaged by living in whirls of words, ideas, images, apprehensions and joys that are partly beyond his or her ken, but, hopefully, no one is left quite the same. It "would not deserve the interest we take in it if it did not have something to teach us that we could not know by ourselves" (Gadamer 1989, p. xxxv) but it would not arouse our interest if it didn't hint at some already operating intimacy. Otherwise, we would not be "addressed" (the locale, Gadamer suggests [1989, p. 299], where "understanding begins"—understanding begins with the sometimes terrible, sometimes exhilarating, sense that *something has already begun*).

It is precisely this "beyond the wall" that is the tale's lure, that provides the possible venture-step into crackling and shimmering surrounds beyond castle walls, or that allows that smoke spiral, *maybe*, to have its consequence.

Maybe it just adds to the sense of foreboding.

Maybe it is key.

Maybe it is nothing.

It may be.

> "People are going to say, 'Well, it's not very truthful,'" says Dylan. "But a songwriter doesn't care about what's truthful. What he cares about is what should've happened, what could've happened. That's its own kind of truth." (Gilmore 2012).

I have had both long and beautiful conversations with very young children about the familiar experience of foreboding that they often know with a secret intimacy and clarity, spots where something *could* happen, and its could-happening floods into our sitting here, some possible future present and full and yet, in its "could," also empty of the sort of specificity that would let it start to rest. Time, here, becomes elastic and pulls the future too close, too taut and taunting, keeping it hidden and approaching all at once. This taut time is one of the "wild things" (Sendak 1988), a *familiaris* (Hillman 2005, p. 107)—an animal spirit (a "familiar," if you will, like the black cat on a witch's broomstick) that can "reveal ... the special quality and dangers ... of a place. To know a situation one needs to sense what lurks in it" (p. 107). "Something is going on, (*im Spiele ist*), something is happening (*sich abspielt*)" (Gadamer 1989, p. 104). This is why Fideles and Old Thomas are, shall we say, familiar figures.

So even in this children's tale, we, too, as adults, suspected something about the lurks of familiarity, of *familiaris*, without perhaps knowing it. We've been here before and yet this is the first time. Someone is already here and it might be me in aspects I've never previously suspected, perhaps ones I've secretly dreaded, perhaps ones I've longed for all my life and always hoped would prove to be the truth of my living ("its own kind of truth" that doesn't describe what is as much as what could have been, what should have been, what might still be). Recognition:

> We do not understand what recognition is in its profoundest nature if we only regard it as knowing something again that we know already—i.e., what is familiar is recognized again. The joy of recognition is rather the joy of knowing *more*

> than is already familiar. In recognition what we know emerges, as if illuminated, from all the contingent and variable circumstances that condition it. (Gadamer 1989, p. 114).

This is how stories and their telling work—they are the open cast of path out into what is both familiar and, at the same time, yet to be known. "Tales lend permission to the listeners" (Yolen 1988, p. 9), but we teachers know a secret. They lend permission to us as tellers as well. We, too, *love* them when they're good, love what they do when they work their magic, love what they promise and can affect. Part of what draws students to stories is the love of a good story, but also the love of a good storyteller who slows time and condenses it in just the proper measure: "That trick and aunt or uncle'd use of always stopping right at the best part to take a bite of pie, a sip of tea, their way of leaning back to look around the table, let the story sink right in" (Wallace 1987, p. 13). The storyteller loves the story, the tells and the listenings. In fact, these are, in the weird whiling arc of story time, the same.

Good stories thus can give us verdant "free spaces," "possible ways of shaping our lives" (Gadamer 1986, p. 59) in which our students (and teachers, too) come back to us from those hills to the south with their own stories to tell of the story they have heard and the smoke spirals they've whiffed. They are a chance, even more intimately, to see how our lives, too, have *already been shaped* in many ways, often "beyond our wanting and doing" (Gadamer 1989, p. xxviii). We find hidden parts of ourselves in them. They can affect us even if we have never cultivated an awareness of them. As an aside, this is why hermeneutics is never simply a report on "my experiences," why it lends itself to and relies upon phenomenology but phenomenology is not enough. The story told must be *lebensweltlich* (the German term for "life-worldly" or "close to the living world"; see Ross, S. 2006, Grondin 2003, p. 333), but the life-world faces us, not with an array of given presences to be simply phenomenologically described. Rather, it arrives as "a task for consciousness and an achievement that is demanded of it" (Gadamer 1989, p.127), "a task that is never entirely finished" (Gadamer 1989, p. 301). It faces us as faces yet to be read. This task of living a life in the life of the world is one whose contours are often occluded, absent, portending, lying, anticipatory, forgotten, suppressed, and we each feel our own deeply embodied culpability in precisely this mess. The life-world, then, is not available for simple phenomenological description. This does not make hermeneutics distant or abstracted from "lived experience" and its intimacies. Far from it. It means that understanding our lives and our way through the world must always be risked, must always be *undertaken*, not just "disinterestedly" documented by a "disinterested spectator" (Husserl 1970, p. 157). Not only is it "everyone's task" (Gadamer

1986, p. 59) to work through how to live with the occluded text(ure)s I've been drawn into and that has already drawn parts of my life. Hermeneutics describes my own intimate experience with the story that is unfurling as a task that no one can undertake for me, instead of me or on my behalf. It is a venture I must take on myself. It is "not something anyone can be spared" (Gadamer 1989, p. 356). Or, as goes the new *de rigueur* in our local school board, keys to the pedagogical effectiveness and quality are "*engagement*" and "*personalization.*" In fact, and here is the terrible rub: even students who "dis-engage" are engaged in a venture that is formative of who they will become, and also formative of who we, as their teachers, become in the face of their dis-engagement (more on this story below, and the tale of F. W. Taylor that is a secret yet intimate part of contemporary education).

So, stories can define us in some small way in our very attempt to unearth their origins and travels. *I myself* am part of the fabric being woven by a loom not just held in my hands alone. "The matches were put into his hand"—every tale, then, a bit of a ghost story rattled full of ancient bones, full of a time beyond the telling.

After all, I can't quite remember how very, very old I was when it dawned on me why dogs might be called Fido—an ancient Roman remnant in English memory, like cows being called Bossy. And then there's the story of how the barnyard is full of Saxon sheep, cows and pigs, and the masters' tables teemed with Normand mutton, beef and pork. Another story, and another, already at work, already at work again, anew, familiar and fresh at the same time. Happened, could have, should have. It seems that we are "*always already everywhere inhabited*" (Smith 2006a, p. xxiv). Part of the work of pedagogy is gathering up our inhabitations, coming to know them, coming to have some small hand in them and their possibilities. Pedagogy has an eye to freeing us from our already-everywhere sleepiness (try reading this statement by Sheila Ross [2006, p. 118] about Gadamer's hermeneutics in light of Bob Dylan's words cited above: "The antonym of truth ... is not the untruth of falsity, but the untruth of utter familiarity"—one might say, a familiarity that has lost its *familiaris*, its animal beckoning). We mustn't be too quick or happy here to simply imagine each student caught in a romance held at the heart of teaching, a sleeping beauty to be awakened. We teachers know that sometimes *we* are the sleepyheads and our students are the awakeners. Their contempt for our sleepy familiarity thus makes them seem wild—that other sniff of *familiaris* that interrupts sleep. Such sudden reversals of fortune are themselves age old stories. Either way, or both, in weaving these tales, we are freed for the task of facing, shaping, remaking, rescuing, embracing, transforming, or fighting off, tearing up and casting aside the possibilities we already inhabit and that already inhabit us. Odd that such freeing ends up entailing a sense of obligation and necessity far outside the confines of schooling. "Something awakens" (Gadamer 2001, p. 50).

So, then, we don't just come to know these stories. We, each in our own way and to our own limits, *recognize ourselves in them*—what is, what could have been, what should have been. Reverse this. These stories *recognize us.* We are *known by them* (see Abram 1996, Palmer 1993). That is why, when the telling works, we sit still and rapt. Addressed. And that is why these matters occur and recur, no story ever quite finished with us. I think, now, writing this sentence, December 16th, 2012, of tales being told of how, yesterday, a Kindergarten teacher, Janet Vollmer, read stories to her children while violence came down outside the closed doorway, in Connecticut (see Carter 2012):

> "If they started crying, I would take their face and tell them, 'It's going to be OK.' I wanted that to be the last thing they heard, not the gunfire in the hall."

And this occurs, such that a story already cited comes round in an arc of recitation and sounds different to the ear, yet oddly the same, oddly, again, *familiar*: "Story is our wall against the dark. They heard the rush of wind, the terrible voice, a scream, then another. Beyond the wall, the dragon waited but could not get in" (Yolen 1998), not this time at least, not, by terrible happenstance, with this class, these children, this terrible chanciness part of what we now must understand all over again, here, in these "stubborn particulars of grace" (Wallace 1987) and grieving.

Time *halts*, and often breath along with it ("*aesthesis*, which means at root a breathing in or taking in of the world, the gasp, 'aha,' the 'uh' of the breath in wonder, shock, amazement, and aesthetic response" [Hillman 2006, p. 36]).

"Shush. Be still." (We've been spotted!)

What will Old Thomas do? (One child has a cat at home that sneaks around. Another hears them yowling, wild things in the night, and stays awake in fear. Yet another asks about that most mysterious thing: purring!).

Is Thomas' switching tail a tell? (A dog's wag is different than this, one child has heard).

And just who is at the door? (That Grade One conversation years ago about keeping the classroom door open, about invitations, welcomes, and what a closed door can do—safety, privacy, or some message of "stay out"—and how not knowing makes the knock worse. "It's the Queen!" "It's the Prince!" "Answer it." "Don't answer it." All this resounds all over again, now, with these latest tales of doors and hallways and horrible unwelcomeness).

Insiders. Outsiders. Thresholds. Guests. Arrivals. Hesitancy. Suspicion. The giggly wide-eyed "uh" as we huddle closer in common breath, conspiratorial (Illich 1998).

Might we best just join the singing fields and leave the sleeping Queen and worried King to their own troubles? Breath deep outside the walls? Or huddle in closer away from the violent dragon's snorffling?

Ah! There. The harsh school-buzzer-bell that breaks the story's gathering, its impending time and its broadening spell with well-measured timeliness.

Out, then, for now, into the sun's recesses whilst another tale is told.

A Familiar Story That Is Not Well-Known: "Time Is Always Running Out"

> Perhaps it is only when we focus our minds on our machines that time seems short. Time is always running out for machines. They shorten our work ... by simplifying it and speeding it up, but our work perishes quickly. (Berry 1983, p. 76).

"By choosing this or that story to tell, I reveal much about myself" (Yolen 1988, p. 13). But in such choosing we also can reveal much about our circumstances as teachers and students inhabiting the haunts of schools.

It is beyond doubt that teachers and students alike experience this phenomenon of time always running out with great intimacy and regularity. Teachers and students alike have become accustomed to the mood, tempo and consequences, personal and pedagogical, of how attempts to try to keep up with this time that is always running out, seem, in the end and seemingly inevitably, to give us less and less time:

> "Well, in our country," said Alice, still panting a little, "you'd generally get to somewhere else—if you run very fast for a long time, as we've been doing."
>
> "A slow sort of country!" said the Queen. "Now, here, you see, it takes all the running you can do, to keep in the same place. If you want to get somewhere else, you must run at least twice as fast as that!" (Carroll 1871, p. 16).

An old story, this. In some schools, this clockwork "machine time," like a demanding Red Queen, seems to render classroom experiences to its relentless demands, pressing itself in on what we do, how we think and imagine, even whether there is time to think much at all. There is almost too much to consider in this orbit—market-driven obsolescence ("our work perishes quickly"), flickering attention spans ("speeding it up") and how such spans then create a world ("simplifying it") that does not *require* much attention, thus aggravating this circle of consequence. Once this voracious and insatiable wheel starts turning, something else kicks in: as the Red Queen suggested, the only relief or fulfillment of this itch is to be found, not in mere *speed* but in *acceleration* (Jardine 2012e).

This suffering is real and palpable. It has become a familiar story—laments about time and its running—both inside and outside of schools, to the extent that talk of any other sense of time, of whiling time and the gathering that happens around good stories, good work, well sought inquiry, and thoughtfulness, seems, in the life of "real world" schools, simply fanciful, unreal almost:

> To be glib, [in this "real world"] little requires human application, so little cultivates it. Long alienated from abiding in inquiry as a form of life and way of being, a restless humanity defers to models, systems, operations, procedures, the ready-made strategic plan, and first and last to reified concepts, long impervious to deconstruction. (Ross S. 2006, p. 111).

Stories of dogs and cats and kings and queens and luxuriating in the migratory arcs of their telling, slowing down and pondering—this experience of time now seems the wonderland. Countless teachers have told me this: they would love to do this, but they simply don't have time, they are always already late no matter what they do, no matter how they try.

This is the core of a sort of ontological delusion that sets in: this sense of relentless, perpetually running-out machine time has become so ubiquitous that it becomes experienced as if it is simply "the way things are." As I hear so often from so many teachers, including those who wish it otherwise, this is simply "the real world." This is where the real perniciousness lies, because once codified as simply "the real world," any attempts to interrupt this spell and suggest that there is a life to pedagogy out from under this ontological delusion are looked upon with great suspicion, accusations of not understanding what it is like, so goes the telling phrase, "in the trenches."

The trick here, of course, is to remember that this is *not* the real world in some intransigent, ontological sense. Rather, *it is how the world has turned out* and therefore, two things. First, there are causes and conditions that can be untangled that can help us understand something of how and why things turned out like this, thus loosening their grip on our imagination and practice. Second, we can perhaps begin to shift the story being told to one that is more amenable to "abiding in inquiry as a form of life and way of being" (Ross 2006, p. 111). This loosening and shifting re-telling is, of course, perennial and tough and full of heartache and therefore cannot be fully fulfilled here.

So, for now, a small offering: three short stories that might help start decoding this familiar story about time running out.

Story One: Empty Time and a Succession of Nows

> The designation 'empty time' is how [Hans-Georg] Gadamer terms time conceived of as the constant, flowing succession of 'nows' coming from a future and receding into the past. This is time subjugated to quantitative measurement. It

> is 'empty' because measuring time requires a separation of the temporal units which measure from that which is measured; to separate time from its contents is to 'empty' it [Ross here references Gadamer 1970, pp. 342–3]. It is in fact the utility function of measured time—time made available for use—that Gadamer says is at the root of this emptying. (Ross 2006, p. 110).

Once we detach our understanding and experience of time from any substantive thing measured in time, time becomes pictured an empty sequence or stretch. Like this, simple: "We've got two [empty] hours this morning with the kids." This empty time can now be "filled" or "used" as we see fit or as circumstances allow. Time thus emptied becomes imagined as something utile, something "useable." Also, and *because of* this imaginal shift, time becomes understood as something that can be "used up," something that can therefore "run out." Example: when parsing through a complex algebraic puzzle, and looking at the arcs that link quadratics to the physical lay-out of a dam that was built in China on the Yellow River and this linked to cultural meditations on the state of the world and its workings (an easily imaginable example in a High School classroom), "inside" of the material, ideational and imaginal fullness of this work, "inside" the thoughtfulness and exploration that this work requires if it is to be done well, time is *not* running out. This puzzling has its own "indigenous" time, it "takes its own sweet time," as the saying goes. It runs in flurries of interest and concern, it slows over beautiful finds or halts and halters sensing dangerous lures, it gathers and stops and starts in shapes and measures that are longing to be proper to the territories of its inquiry. Oddly put, such work needs our attention and devotion, and its very substantiveness slows our investigation, turns it here and there, and makes demands on the while we take over it. We may not "have" enough time to consider it well, but that does not mean that it doesn't have its own temporal inherence, only that we cannot live up to that behest.

In light of utile, empty, measured-time, those things that we have to do, our work, our teaching and learning, the topics we explore, become no longer understood to have a time of their own. Rather, pedagogy is understood as occurring within specific measures of useable time, empty time. This is where the turn occurs: the work that can then be pursued under such auspices is *rendered measurable* by such empty, formal, clockwork temporality. It is not simply that the things we have to do are molded into a tempo that is fast and efficient. Those very things themselves must, of a necessity borne of this empty time cast in a sequence of "nows," become fragmented into pieces that can fit the measure of empty time itself. *What we have to do* (exploring animals and their habitats in Grade Five, learning about democracy and its characteristics in Grade Ten, or reading about dogs and cats in Grade One, and so on) *changes in order to shape itself to the useable-ness of empty time.* Given the sequenced march of empty time, then, only once things are fragmented into sequence-able bits and pieces can the things we are charged to do (e.g. mandate curriculum objective/topics, etc.) "fit" the ever-accelerating succession of "nows" that empty, measurable, machine-like time demands of things. To the extent that what we do *cannot* thus shape itself, to that extent, we have to eradicate from consideration such pursuits. China, the Yellow River, the quadratics of dam-building, the nature of displacements of water and people, the structure of government decisions. All this is quite lovely, but we don't have time.

Thus, time is not simply *subjugated to quantitative measurement* (we mustn't forget this, however, because that story-telling time of Fideles and his worries has been subjugated and often marginalized in many classrooms to the hurries of skill development. "We'll get to reading a story later if we have time"–the indigenous time of good work gets relegated to what is done "after" that work which is more, as it is called, "basic" [see Jardine, Clifford & Friesen 2008]. More on this below). Empty time now *subjugates* anything to which it is applied and marginalizes anything that cannot be thus subjugated. The thing now measured "in [empty] time" must itself, in its very substance, become the objective equivalent of a series of "nows"—separate, self-contained, isolatable fragments or pieces—that must be then assembled in sequenced, ordered, managed and standardized in order to be adequately temporally measured and, especially, in order to be, as the saying goes "covered" in the allotted time. Thus a hidden logic churns: as things fragment, time accelerates *because* there is nothing to slow it down since no one of these isolated bits or pieces *requires* any prolonged attention. Once detached from things that live in a time proper to those things, empty time produces fragments that no longer *need* "continuity of [our] attention and devotion" (Berry 1986, p. 32). Worse yet, these fragments reject and cause to atrophy that very sense of devotion, making it seem, not surprisingly, like a "waste of time."

Empty time thus now rules the work being done. But it is important to re-emphasize what has happened here. This is not just a matter of demanding that the same work being done be done "faster" (this would be as mistaken as imagining that the "slow food movement" is suggesting cooking fast food slowly). Nor does empty time simply rule *how* the work is to be done. *The very nature of the work itself changes,* as does the relationship that one can strike up with the work. And, to reiterate a point noted above, work "covered" in used-up, empty time work becomes understood as more "basic" than the (now thought to be) luxuriousness of "abiding in inquiry." Worse yet, what might be attended to in a luxurious way—say that story of Fideles and his fidelity, or the wetland down the way from the school that we might visit and "study"—is understood to be "really" made up of pieces and in order to get to the whole of that story, we must have the pieces out of which it is made *beforehand*. We must, as goes the familiar phrase, start with "the basics."

What happens then is also all too familiar. This trumping of empty time is not just a matter of chronological deferral—"we'll get to that later if we have leftover time that hasn't been used up." That which we might get to later—to use the shorthand, "abiding in inquiry"—becomes not simply (possibly) "later," but becomes understood as a "frill" that is unnecessary to the "real world" reality of things. Such matters become leftovers, extras. In the real world of schools and Provincial examinations and parents' demands for "accountability," getting to it "later" is not really an especially urgent manner "in the real world" because the real world is now understood to *be* isolated, testable, controllable, predictable, and manageable fragments.

After all, "first things first."

But here is a school reality that is hard to admit: those sorts of work that fit the clockwork, one-thing-after-the-other, always accelerating rush of empty time bully themselves to the front of the line and provide a way to not just marginalize but humiliate those who might suggest that there is thoughtfulness, rigorousness, authenticity and good work to be had out from under this running-out panic.

In short, bullying in schools really is a very important topic and far more rooted in the very nature of school once it is conceived under the auspices of efficiency.

Story Two: Industrial Production and the Efficiency of Schooling

> "Education is suffering from narration-sickness," says Paulo Freire. It speaks out of a story which was once full of enthusiasm, but now shows itself incapable of a surprise ending. The nausea of narration-sickness comes from having heard enough, of hearing many variations on a theme but no new theme. (Smith 1999, pp. 135–6).

> The uniformity, standardization, and bureaucracy of the factory model soon became predominant characteristics of the school district. The key was to have the thinkers of the organization specify exactly what and how to teach at each grade level and then to provide strict supervision to ensure that teachers did as they were told. Decisions flowed from state boards of education down the ladder of the educational bureaucracy to local school boards, superintendents, and principals. Eventually, decisions would be directed to teachers who, like factory workers, were viewed as underlings responsible for carrying out the decisions of their bosses. Students were simply the raw material transported along the educational assembly line. They would be moved to a station where a teacher would "pour" in mathematics until the bell rang; then they would be moved to

> the next station where another teacher would "assemble" the nuts and bolts of English until the next bell rang, and so on. Those who completed this 13-year trek on the assembly line would emerge as finished products, ready to function efficiently in the industrial world. (DuFour & Eaker 1998).

> Most schooled tasks have been stripped of that character which would take a while. A "continuity of [our] attention and devotion" (Berry 1986, p. 32) to some classroom work is very often not simply *unnecessary* but *impossible* because the school-matters at hand have been stripped of the very memorability and relatedness ... that might require and sustain and reward such attention and devotion. From the point of view of efficiency and management, intellectual whiling in the leisures (*schola*) of school simply seems dense and unproductive. [Little in school tasks organized thus is] worth *while*. (Jardine 2012e, p. 175).

At the beginning of the twentieth century, there was a profound shift in the way in which industrial production was imagined, organized, and carried out, and what was, at first, a brilliant shift, occurred, full of enthusiasm. F. W. Taylor (1856–1915), most explicitly in his still-published text *The Principles of Scientific Management* (1911), instituted what was later to be called "the efficiency movement" (Callahan 1964). This movement arose out of Taylor's observations on the shop floors of various industries on the East Coast of America (Bethlehem Steel, for example) and his development of what he called time and motion studies.

It had been that artisans and workers would gather around the work to be done in ways that were age old and linked to ancient guild and master/apprentice organizations and to the sometimes written, but often oral transmission of knowledge and craft, of hand laid over hand, of breath and bread shared over the immediacies of laboring. Taylor entered this fray as an observer (I'm imagining Husserl's disinterested spectator all over again), and conducted time and motion studies of such industrial production. Essentially, he measured every step of the work being done—who was doing what, what others were doing in the meantime, even literally how many steps and in what direction anyone would take to get materials, to work around other worker or wait for them, and so on. In effect, Taylor temporally ("time") and spatially ("motion") broke down any particular industrial task into its, shall we say, basic, component parts and laid out ways in which to organize, manage and sequence those separate parts more efficiently. He experimented in great and meticulous detail with the sequence of the work, the portioning of the work, the effects of placing accessible parts or tools here or there, with this first instead of that, of this worker doing these three things, or two, or perhaps having this one arduous task done by two workers, turning to the left to grab the next part to be assembled, or to the right, working and resting for these lengths, in this order or that, this amount of training done this way, that way, and on and on, tumbling these components all with an eye to the elimination of waste—wasted time, wasted materials, essentially, wasted money—and all this with an eye to increasing the productivity of the work being done, decreasing the errors and glitches encountered, and increasing, thus, *efficiency*.

All of these studies were traced with stopwatch in hand, notebooks and measuring tape, leading to a new invention: flow-charts for the work being done. As was the atmosphere of the early 20th century, this work of Taylor's was touted with the portentous term "*scientific* management" to contrast it with old, rule of thumb, practically based work—we need to keep our eye on this shift, because this is part of the movement of evacuating from front-line practices any knowledge or worth and placing in the hands of managers/administrators/principals the task of organizing work so that workers need not think, need not be especially "skilled" but only obedient to the system of work devised by management, thus making labor cheaper and thus increasing the efficiency of production as a whole (on the "deskilling" that comes from this movement, see Braverman 1998).

Henry Ford's car assembly line provides us with an easily recognizable image of what was, in fact, a "culmination of a decades-long process" (Watts 2006, p. 153) initiated by Taylor: each worker has placed in front of them an isolated, repeated task to be done with singular, standardized procedures, invariant materials and a specific allotted time and sequence for its completion (on Taylorism and Fordism, see Kanigel 2005, p. 49):

> The basic procedure made management the absolute arbiters of when, at what speed, and in what fashion the work was performed. The assembly line's smooth, continuous flow, in the words of Horace Arnold, worked by "hurrying the slow men, holding the fast men back ... and acting as an all-around adjuster and equalizer." It was the apotheosis of scientific management. (Watts 2006, p. 154).

(I can't help but think, here, of the mathematics department in a local High School that requires not only all those learning, but all those teaching mathematics in Grade Ten to be on the same chapter at the same time for the course of the semester). Even though there is no evidence that Henry Ford actually read or was directly influenced by Taylor's work, "the Ford Motor Company was 'Taylorized without Taylor'" (Watts 2006, p. 153):

> Factory managers struggled to break the hold of artisan craftsmen, with their traditions of stubborn independence, and fought to eradicate ... "premodern" work culture, with its agricultural aversion to disciplined, time-oriented labor. They sought to construct a new model of labor more attuned to the demands of efficiency and mass production. (p. 153).

Note in passing here how the profound cleaving to a deep and well-understood sense of time that is indigenous to the ways of agriculture is simply swept aside in this statement, and how, therefore, the specific sort of "discipline" that comes with empty time casts agriculture as seemingly undisciplined, slack, stubborn—note, too, how "independence" becomes cast as a disparagement.

So, under the auspices of efficiency, all tethers of one specified task to any other tasks or to the object being assembled or any tether between this worker and other workers, or tethers to the ancient arts of craft and work, or tethers to any concern for the quality of what is being done or the purpose—all this has been systematically eradicated as detrimental to the efficiency of the work being done:

> "Every day, year in and year out, each man should ask himself over and over again, two questions," said Taylor in his standard lecture. "First, 'What is the name of the man I am now working for?' And having answered this definitely then 'What does this man want me to do, right now?' Not, 'What ought I to do in the interests of the company I am working for?' Not, 'What are the duties of the position I am filling? Not, 'What did I agree to do when I came here?' Not, 'What should I do for my own best interest?' but plainly and simply, 'What does this man want me to do right now?'" (cited in Boyle 2006).

I should also not ask why we are doing this, what this is part of or leading to, what my role is in all of this. The task for industrial factory workers is simply to learn by rote and repetition the efficient accomplishment of this one, isolated task and then to either simply repeat that task or get on to the next, equally isolated task at hand (one can hear, here, the roots of the laments regarding disengaged students in schools, which is uttered with little knowledge that such disengagement is a *deliberate* and *necessary* feature of the efficiency movement's importation into schooling. Many school reform movements try to address such disengagement while leaving in place this forgotten history and its still-lingering effects). Just again as a side note that rings through still:

> This revolution in the conduct of labor also transformed its soul. For generations, American attitudes about work had been rooted in the Protestant work ethic. The assembly line, by making labor monotonous and unfulfilling, eroded the foundations of the ethic. It raised troubling questions about the meaning of work. (Watts 2006, pp. 154–5).

Henry Ford's solution to this conundrum is too astoundingly familiar to detail here. Suffice it to say this: the, shall we say, "pay-off" for doing monotonous and unfulfilling work is not found in the work, but in its resultant "consumer abundance" (Watts 2006, p. 155). The ethic of work and its fulfilling pleasures is replaced with the ethic of consumption—the pleasure to be had, to be purchased, *after* the work is done:

> Work and play should not be mixed. "When we are at work we ought to be at work. When we are at play we ought to be at play," [Ford] wrote. "When the work is done, then the play can come, but not before." (Watts 2006, p. 155).

The once "playful" pleasures and engagement and fulfillment of good work are evacuated from the work itself (which becomes fragmented, routine and monotonous) and the higher wages then paid for obediently enduring such monotony can be exchanged, afterwards, for purchasable, enjoyable things. Engagement itself becomes a leftover caught in regimes of market-exchange, a great analogy to justifying and enduring the boredom of routinized High School classes in order to receive marks that can *then* be exchanged for future employment that *then* can be exchanged for one's chosen enjoyments. Note the great eschatological arc of empty time here, where the present is drained of its life with the promise of future fulfillment: some time in the future, time will no longer be empty but full. Thus the Protestant work ethic re-emerges in the gruff anger swirling around "this is the real world" talk in High Schools, of the "get used to it" reprimands in response to students' (and teachers') resistance to the monotony. Thus, too, an origin of education's obsession with the idea of "time of task."

As a result of F. W. Taylor's re-imagining of industrial production, industrial efficiency increased dramatically. Moreover, this image of efficiency and its promise took over the public imagination and swept through all facets of then-contemporary life, from mayor's offices to hospitals to how housewives should organize their kitchens and their housework schedules and on and on (for more detail on these matters see see Taylor 1903, 1911, Kanigel 2005, DuFour & Eaker 1998, Callahan 1964, Gatto 2006, Wrege & Greenwood 1991, Friesen & Jardine 2009, to name but a few available sources). Dozens of articles in popular magazines and scholarly journals were written and poured over, along with recurrent declamatory newspaper articles about the inefficiencies of this or that facet of then-contemporary life. "What about efficiency?" became a polemical, even moral clarion call in all quarters of North American consciousness: "Taylor's thinking so permeates the soil of modern life we no longer realize it's there. It has become, as Edward Eyre Hunt, an aide to future President Herbert Hoover, could grandly declaim in 1924, 'part of our moral inheritance'" (Kanigel 2005, p. 7).

In reference to Hans-Georg Gadamer's explorations of empty time that is always running out, Sheila Ross (2006, p. 118) suggests that "the dominance of this modality of thought ... is arguably pathological." Just as a reminder, it is not that this way of thinking is pathological. It is its *dominance* that is at issue here, and how easily it has come to occlude, bully and marginalize other ways of experiencing work, time, engagement, learning, and so on. It has lost its sense of proportion and place due, in fact, to its being *premised on* the fragmentation of any territory it enters and considers. It thus cannot be expected to find its own limit. It is, thus, pathological in its sway.

Given the burgeoning numbers of immigrant children entering large East Coast American cities, and the equally burgeoning need for minimally educated workers in industry, schools had become overwhelmed early in the 20th century, and the promise of more efficient schooling was irresistible: "educators needed little prompting" (DuFour & Eaker 1998). Thus we hear from Ellwood P. Cubberley, Dean of the School of Education at Stanford, from his book *Public School Administration*, originally published in 1916 [cited here from Callahan 1964, p. 97]):

> In time it will be possible for any school system to maintain a continuous survey of all of the different phases of its work, through tests made by its corps of efficiency experts, and to detect weak points in its work almost as soon as they appear. Every manufacturing establishment that turns out a standard product or series of products of any kind maintains a force of efficiency experts to study methods of procedures and to measure and test the output of its works. Such men ... [also] train the workmen to produce a larger and a better output. Our schools are in a sense factories in which raw products (children) are to be shaped and fashioned into products to meet the various demands of life. The specifications for manufacturing come from the demands of twentieth-century civilization, and it is the business of the school to build its pupils according to the specifications laid down.

There is, of course, much, much more to this story of the re-capitulation of empty time sequences in the shape of industrial production and how these images found a great ally in a then-emerging theory of knowledge which imagined knowledge as built out of separate bits and pieces:

> As behaviorism grew in prominence during the period between the two world wars, the scientific management movement in education was being promoted by [Franklin] Bobbitt (1924), a curriculum specialist who, citing the need for efficiency and using the steel industry as his model, attempted to apply the techniques of business to the schools. In the name of efficiency, he gave paramount importance to the setting of acceptable performance standards and to their measurement. He formulated long and detailed lists of objectives which he felt would enable learners to prepare for life by mastering specific skills and subskills. (Tumposky 1984, p. 296).

Part of the attractiveness of Taylor's industrial promise of efficiency thus dovetailed with the logic of fragmentation borne from a since-outdated version of the empirical sciences in the early 20th century. The then-emerging Behavioral Sciences produced an image of knowledge as built up one "basic" bit at a time. Each separate fragment is itself and has only "revocable and provisional" (Gray 2001, pp. 35–6) connections to anything else. "The basics," in education, became identified with those not-further-divisible "bits" out of which any knowledge was built, and "back to the basics" (see Jardine, Clifford & Friesen 2008) comes to mean back to a version of knowledge-assembly right in line with Taylor's industrial-assembly principles. Taylor's influence thus found broad affiliations when applied to education, and its influence filtered down into the very ways in which the topics covered in schools were imagined to exist: sequences of separate parts and therefore, through the grades, assignable parts of such assembly to be "covered" in each grade. It may be that classrooms prior to such infiltration were already cast this way, but Taylor's work, coupled with Behavioral Theory, gave this cast an air of modernity, of seriousness and scientific warrant that provide a way to trump any resistance to its influence.

It also provided a clarity and distinctness that resonated with a certain unnameable but secretly influential, almost mythological links (shades, too, of Freudian squirms): "The unclear is the unclean," (Turner 1987, p. 7), thus linked the "decontaminated" bits and pieces on Taylor's assembly line with the clarity with which they can be scientifically managed.

It is impossible to detail all this in the present context. Instead, then, just a few closing bits and pieces, trails that can be followed, half-forgotten stories that can be hunted out, filled out, and re-told:

- From Henry Ford's (2007, p. 14) autobiography: "eliminate the useless parts. This applies to everything—a shoe, a dress, a house, a piece of machinery, an airplane, a steamship." Following the inculcation of Taylorism into education, this applies to each and every minute spent in the classroom, each and every topic that is learned, how schools and school departments are organized, and how the work of students, teachers and administrators are apportioned. Eliminate useless parts as well as useless replications of time and motion and effort, where, in each case, "uselessness" is defined in advance as that which cannot be efficiently learned, efficiently tested, and accounted for. What happens then is that, for example, heated conversations in a Grade Ten classroom about the exportation of democracy to other countries via pre-emptive, invasive actions are deemed "useless" in light of standardized assessment regimes that will be testing for students' ability to name four characteristics of democracy on upcoming Provincial Examinations. Those conversations become a waste of time, a, so to speak, "luxury item" that can be purchased *after* the monotonous memorization of those characteristics, providing that memorization is done efficiently enough to "save" enough time for such luxuries.
- The product thus efficiently and repeatedly produced becomes effectively identical every time. "Any customer can have a car painted any

color that he wants so long as it is black" (Ford 2007, p. 72). One can read stories about Fideles and Old Thomas "so long as" reading skills are secured beforehand. And, of course, it is no longer the *responsibility* of schools to pursue such readings. They are "frills" unless such readings are accountably linked to outcomes such as "attends to story structure" or "is able to predict based on previous information." Because the desired product is thus standardized, a standardized assessment of the results of production can be developed and applied uniformly and without variation. To the extent that our "readings" of this story cannot be rendered identical, those readings are rendered "subjective," thus, again, marginalizing multifariousness in favor of self-sameness, favoring singularity and identity over diversity and abundance.

- Not only is the *product* standardized. Efficiency requires the "complete standardization of all details and methods. [It] is not only desirable but absolutely indispensable as a preliminary to specifying the time in which each operation shall be done, and then insisting that it shall be done within the time allowed" (Taylor 1903). There is thus only "one best way" (Kanigel 2005) for *what* is being done, *how* it is being done, *how long each step takes* to be done and the precise and undeviating *order* of such steps. Moreover, exactly *who* is responsible for *what* can also be specified with precisely the same standardization and uniformity. Anything that now deviates from this is considered an error in the system that needs eradication. Luckily, because of the standardization, the source of error is easy to find. For example, if we have a child who does not understand "place-value," we can specify precisely where "on the line" this understanding was to be "built," precisely where on the line one would have tested to ensure such building (or ensured that this line-step was repeated until such assurance was had) and so on. Of course, there may not be a defect "in the system." There may, rather, be a defect in, to use Elwood Cubberley's phrase, "the raw product"—the child may be "defective." And as those of us in education know full well, such possible "defects" have lead to astoundingly complex regimes of alternate assembly lines for everything from "the retarded" to "laggards in our schools" (Ayres 1909), to children with special needs, to learning delays, slow children, Individual Program Plans (ready made and individually designed assembly lines geared to the precise "raw product" deficits [see Gilham 2012, 2012a]). Thus, even though, Taylor insists "in the past the man has been first; in the future the system must be first," (1911, p. 2) the school system leaves the fragmentation and sequencing of Taylorism in place and then simply starts, on behalf of the well being of individual students and their diversity and difference, to multiply that system into sub-assembly lines. Any student unable to meet the standardized demands of the main assembly line is thus ejected from than line, purportedly for their own good and to accommodate their special needs, but there is another reason, too. Such ejection is the most efficient way to maintain the efficiency of the main line of schooling. Difference, deviation, resistance, interruption, questioning too much, not obeying or being adequately compliant—all of these things are not taken to be indicators, perhaps, of something about the line itself, but pathological indicators of "abnormal" student characteristics.

- From a June 4th, 1906 lecture by F. W. Taylor (cited in Kanigel 2005, p. 169): "In our scheme we do not ask for the initiative of our men. We do not want any initiative. All we want of them is to obey the orders we give them, do what we say, and do it quickly." Recall "little requires human application, so little cultivates it" (Ross 2006, p. 111). Initiative and interest become cast as a *detriment to efficiency itself.* They are, again, luxury items, leftovers. And students learn this quickly and become complicit in this logic. The student who asks too many questions becomes "taught a lesson" later in the hallway. The student who appears interested in what is being learned becomes marginalized as a "teacher's pet." Likewise, the new teacher who shows enthusiasm for their work is tolerated, but inundated with war stories about how the experienced teachers in the school, too, in their day, were once full of enthusiasm, interest and initiative. Thus, the restless student becomes named as suffering from Oppositional Defiance Disorder in the same gesture that names the new enthusiastic teacher green and naive. In light of the smooth operation of the line, "independence and stubbornness" become seen as a little "wild" and "undisciplined" while "discipline" now comes to be equated no longer with a hard-won practice and involvement, but with the compliant following of rules, and all that in a system now hysterically aroused over issues of bullying.
- Taylor's "declared purpose was to take all control from the hands of the workman (whom he regularly compared to oxen or horses) and place it in those of management" (Kanigel 2005, p. 19). "What [Taylor] really wanted working men to be [is] focused, uncomplicated and compliant" (Boyle 2006). Educational psychology then often unwittingly conspires with theories of normality and abnormality, deviance and resistance, theories of giftedness and slowness, in order to codify these purposes.

- Since effective schooling becomes linked with the obedient following of rules and following them the way anyone and everyone follows them, it becomes clear that to be a student *and* to be a teacher means to be *completely and utterly replaceable*: This is part of the "scientific" character of "scientific management," one of the consequences of how it operates. To the extent that who is learning and who is teaching makes a difference in the classroom, to that extent, that teaching has become, as goes the term in a failed experiment, *contaminated*. Any innovative practices are thus marginalized as the gifts of a particular teacher, thus once again preventing the normalcy of efficient schooling that can be done by anyone from being interrupted—one might easily say "contaminated"—by such examples.
- Given this link of contamination with inefficiency, consider that H. Martyn Hart, the Dean of St. John's Cathedral in Denver, from the September 1912 issue of the *Ladies' Home Journal* decried decreases in character, increases in divorce and crime, lack of self-control, illicit political machinations and attributed all of this to *inefficiency* in "the system of schooling" (Cited in Callahan 1964, p. 52). Simply consider here how contemporary schools have become "hot spots" in this regard, where economic, social, cultural, political, and moral praise and blame swirl.
- What then happens is a sort of "ghost echo." This tangled story of efficiency and its insinuation into education is not just a story produced in a closed system. It is story and a system dedicated to *producing* closure. Any attempt to interrupt this story can be cast aside without hesitation because that attempt can now only be speaking on behalf of *inefficiency*. After declaring that "the man" used to be first but now "the system" is first, Taylor goes on to write that "this in no sense implies that great men are not needed. On the contrary, the first object of any good system must be that of developing first-class men; and under systematic management the best man rises to the top more certainly and more rapidly than every before" (1911, p. 2). Remember, though, that "greatness" and "first-class" here mean those who can maintain the system and its efficiencies. Taylor's purpose in the introduction of the suggestion box in such settings is to take suggestions that, over time, *eliminate the need for further suggestions*—this is the moment where the system becomes first, and "the man" is simply either fitting into it or ejected out of it (teachers and students alike). This is the same "closed circle" echo found in *The Fraser Institute's Annual Report Card on Alberta Schools*: "If teachers were following the provincial curriculum by

definition they would be teaching to the test. If they're not teaching to the test, then they're not doing their job" (McGinnis 2008, p. B5, citing the institute's Peter Cowley). Once this loop closes, the dominance of this form of thinking then projects upon any dissent the character of being irresponsible, being "unaccountable." "Abiding in inquiry" looks like letting kids do whatever they want and to hell with the curriculum, let's just be free and arty and "creative." Thus rear up again those images of wildness, of the threat to the well-organized Capital of the stubborn and independent life of the fields (in Latin, *paganus*, root of the term "pagan"; Old English heath, root of the term "heathen"—in both cases, related to those who work the fields):

> Restless students [ones asking questions beyond the efficient ken of the efficient classroom, ones who ask where we are, what we are doing here, why we are doing these things or who are resistant to be considered replaceable, or who refuse to be disinterested or who text each other at the back of the class] become portends of utter, unutterable Chaos. As I've witnessed in many High Schools, there is the unuttered belief that if you let go for a minute of the narrowed and fenced regimes of management and control, quite literally, *all Hell will break loose*. That this hallucinatory vision of the threatening Hellishness portended by restlessness is, in some part and however unintentionally, *produced by* the very narrowing that has been set up to protect us from such a threat—this becomes too horrible to contemplate. (Jardine 2012a, p. 83).

- Finally, schooling, *of all things*, becomes rife with anti-intellectualism (see Callahan 1964, p. 8).

One last thing in this lengthening short story. Looping back to our considerations of time, and weaving this with our contemporary concerns in Canadian schools with the burgeoning multicultural face of our classrooms, consider this brief reminiscence from "William C. Klann, foreman of motor assembly at the Highland Park facility" (Watts 2006, p. 142): "One [phrase] every foreman had to learn in English, German, Polish, and Italian [and now Hindi, Urdu, Arabic, Mandarin and others] was 'hurry up'" (cited in Watts 2006, p. 154).

Story Three: On Spelling Lists and Kings and Queens

One last little story in honor of Fideles.

So we have an admixture of industrial production based on fragmentation, standardized sequencing and the requirements of obedience and the tolerance of monotony for a promised eventual payoff, linked with a Behavioral theory of knowledge that considers knowledge as something likewise built up of fragments and pieces. Both of these lead to a pedagogy premised on breakdown, giving the classroom an air of an early 20th century experiment, where contamination has been controlled for, and where the living knowledge of the living world is rendered into an artifact to be assembled. Thus Gadamer (1989, p. 336) speaks of "the ideal of knowledge familiar from natural science [especially, not coincidentally, from the version of natural science taking hold right at the time of Taylor's work in industrial assembly], whereby we understand a process only when we can bring it about artificially." And thus we get a description of such an ideal that is perfectly in line (pun intended) with Taylor's imagines of industrial assembly:

> The object is disassembled, the rules of its functioning are ascertained, and then it is reconstructed according to those rules; so, also, knowledge is analyzed, its rules are determined, and finally it is redeployed as method. The purpose of both remedies is to prevent [preempt, one might say] unanticipated future breakdowns by means of breaking down even further the flawed entity and then synthesizing it artificially. (Weinsheimer 1985, p. 6).

It is only half a step, then, to the work of Franklin Bobbitt (1918, 1924) and Werrett Charters (1923); one hop from there to Ralph Tyler (1949), and one more to the underpinnings of No Child Left Behind. And we can see here another familiar phenomenon. The *more* difficulty a child has with learning some isolated fragment of knowledge, the *more* we break it down, the simpler and stupider and more demeaning the task gets, the less initiative required, the more sheerness is the required obedience. After all, children need "the basics." Very young children, then, are treated in very recognizable ways under this auspice, as the front end of a controlled Behavioural experiment in the production of knowledge in ways akin to the efficient production of an industrially produced and reproduced object, each one identical to the others both in the process of production and its allowable outcomes.

So one last short story before the children come in from recess (remember? It seems a long time ago we were rapt by Fideles and the King's worries and that intensely leisurely time of engagement). Initiatives parallel to the burgeoning efficiency movement and its behavioral consorts were accomplished in the area of language by Leonard Ayres in his *A Measuring Scale for Ability in Spelling* (1915) (see also his *Laggards in Our Schools* [1909], the full text of which is available on-line).

Through various experimental means, Ayres determined the 1000 most *frequently used words* in the English language. Already we can see that the culling of the "field" of the living language is done according to a quantitative measure which makes it manageable and survey-able in a very particular way: frequency already considers words to be what they are independently of the tales or landscapes that surround them. Surroundings are ambiguous, overlapping, insinuating, localizing, and so on. Language is thus "subjugated to quantitative measurement" (Ross 2006, p. 110) (and thereby the subjugations of empty time and its running) which demands that the insinuations, invocations and tethers of one word to another are no longer taken to be relevant or in force. "Frequency of usage" provides a list of *separate*, self-contained "objects" that can be subjected to efficient, standardized, sequenced, artificial assembly.

Ayres then tested groups of students of gradually and carefully measured increasing age/grade level on their ability to spell correctly such words. Individual words were then tested to see if children recognized them in isolation and the statistical frequency of errors made by children of a certain age were measured. As older and older children were tested, the frequency of errors in the recognition of a given cluster of isolated words decreased. Once the frequency of errors reached a certain threshold for certain words, a line was, so to speak, drawn in the list of words. When a certain percentage of a certain age of students consistently got certain words *incorrect*, Ayres, so to speak, drew a grade-line that indicated that they were "harder" words to recognize or spell, and therefore learning them belonged in the "next" grade. When a certain percentage of a certain age of student consistently got certain words *correct*, those words needed to be learned in a lower grade in school. In this way, spelling lists composed of isolated words for various grades were produced and continue to be produced, and literally hundreds of these are available on-line, including Ayres' original lists (See Rodgers 1983, 1984).

In Ayres' work, not unlike F. W. Taylor's mode of operation, any lingering tethers of sense and significance *between* words being learned by children were ignored or eradicated, not because such tethers didn't exist, but because, in a system of efficient assembly, such tethers constitute "contamination." Words and their allures thus become "terms" whose terminalities have been set in advance of any *post hoc*, illicit relations (words, weirdly, come "after" letters, and sentences after words, and stories after that). First children have to learn the basics. "It," "is," "the" then take on the oddest of empty time characteristics—they are to be learned *before* (and this even though the child comes to school spelling Tyrannosaurus with great ease—this is not standard issue, and therefore irrelevant).

Students and teachers thus immediately experience a cascading logic similar to the panics of the industrial assembly line. The detachments or severances of one word from another, of one word from the story in which it belongs, of the tethers that bind students to the tale being told, of the tethers that bind the teacher to the story, of the tethers that then bind the teacher (and parents as well) to the student *in* the telling territory of the story—cutting all these tethers is a means of ensuring surveillance, standardization, and efficiency of assembly. But then, such detachments and severances make us all feel "cut off," even though it is difficult to remember precisely what it is we've been cut off *from*. Teachers and students and parents and administrators thus start experiencing what is at first a low-level sense of unease, of time running out, of not keeping up, of not quite knowing where you are at any turn because no one bit of work invokes any sense of place or locale. (I can't help think of the relatively recent shift from phone conversations that begin "Hi, how are you?" to "Hi, *where* are you?" This may or may not turn out to be a version of the same story being told here).

Then the turn around again: efficiency and surveillance offer themselves as a *remedy* to this growing unease, and the lingering around stories of cats and betrayals and smoke spirals is cast as the increasingly suspicious *cause* of our unease. "Here, use these already color-coded developmental readers instead of getting lost in a complex story that doesn't allow easy tracing of where each child is 'at.' They will allow you to put your mind at ease, knowing that the course towards literacy has already been totally mapped out in advance of any particular child's arrival, and all possible futures are already foreseen." Thus David G. Smith's (2006 p. 25) horrifying insight about the temporality of such matters, a sort of "frozen futurism": "There *is* no future because the future *already* is," a terrible sentence that couples how efficiency forecloses on the future while, at the same time, portends weirdly apocalyptic fears and anxieties. Hence the need for such a short story, because what gets hidden here is how the promise of efficiency is first produced out of a fear of the unforeseeableness of the future, and then produces the very detachments and consequent unease for which it then offers itself as the only scientifically warranted and manageable and accountable cure .

But these English words "king" and "queen" belong together, and this with all the cultural specificities involved, all the Eurocentrism, all the threads and lines and lineages, all the invocations of images of power and exclusion and realms and influence and war, all the weird violence perpetrated under the Crown in all its myriad tellings. There is a site called *Time 4 Learning* (www.time4learning.com) that has spelling lists available to print by Grade Level. On the page listing spelling words for Grade One students we find the word "king." (http://www.time4learning.com/SpellingWords/1st-grade-spelling-words.shtml). Another lists spelling words for Grade Three students include the word "queen" (http://www.time4learning.com/SpellingWords/3rd-grade-spelling-words.shtml).

A site called *BigIQkids* (www.bigiqkids.com) lists "king" in Grade Two (http://www.bigiqkids.com/SpellingVocabulary/Lessons/wordlistSpellingSecondGrade.shtml) and does not list "queen" at all. *AAASpell.com* lists "king" in Grade One (http://www.aaaspell.com/grade1.htm), also with no listing for "queen." In a word-bank list of the 1200 most high-frequency "writing words," "king" ranks 705th (http://school.elps.k12.mi.us/donley/classrooms/berry/sitton_spelling_activities/4thgrade_spelling/sitton_word_list.htm) and does not list queen in the top 1200. The *Fry List* has "king" ranked as the 490th most frequently occurring sight word and "queen" is not ranked (www.cantonschools.org/content/pdf.../high_frequency_words.pdf). The infamous 1915 *Ayres Spelling Scale* (http://www.ipmpro.com/item/455/Ayres+Spelling+Scale) lists neither "king" nor "queen" in the 1000 most commonly used words.

Dare say, I could go on.

"Time Is a Bringer of Gifts"

> Time is a bringer of gifts. These gifts may be welcomed and cared for. To some extent they may be expected. They cannot in the usual sense be made. Only in the short term of industrial accounting can they be thought simply earnable. Over the real length of human time, to be earned they must be deserved. (Berry 1983, p. 77).

Now where were we when the bell rang? It is vital, as we turn back towards a richer experience of time—what H.G. Gadamer calls "full-filled" time in contrast to "empty time"—that we don't unwittingly drag with us entrails of these short stories we are attempting to leave behind. An all-too-common critique of and resistance to the empty rush of Taylorism in schools is to *leave the bits and pieces in place* and *leave in place the underlying assumption of empty time*, and simply de-standardize the assembly. In what amounts to nothing more than an inversion of Taylor's (1911, p. 2) "in the past the man has been first; in the future the system must be first," each individual student is put "first" in a fragmented and scattershot world, each making what they will of the bits and pieces. We end up with assembly lines democratized, now plural and diverse, minus surveillance and uniformity. This, of course, ends up simply multiplying the rush into a sort of post-modern proliferation and scattering, leading, seemingly inevitably, to reactionary backlashes and new invocations of "back to the basics" as a tried and true remedy and fail-safe means to accountability. Thus we seem stuck with an old, worn-out options, between the closures and foreclosures of "the system" or wild, scattershot, undisciplined chaos. We lose sight of the fact that these feared images of wild, scattershot, undisciplined chaos are, in part, precisely a *product of* precisely the system which deems "all Hell breaking loose" as the only alternative to its machinations.

Breaking out of these old, worn-out, self-perpetuating options is tough work because they form the very atmosphere of our attempts to break out of them. Going back to Wendell Berry's invocation of time as a bringer of gifts (recall that Berry is a *farmer*, one of those whose "stubborn independence" Taylorism and Fordism "fought to eradicate" [Watts 2006, p. 153]), the trouble is, for the gifts that full-filled time can bring to be deserved *requires something* of those anticipating such gifts. We need to *do something* more than simply demand, "Tell me exactly how do you do it?" We have to be adamantly wary that this sort of demand, and our assumption that the forthcoming answer should be simple and not require much of me, are *precisely* how Taylorism continues to lurk in efforts to step away from it. The worthwhile time of abiding in inquiry and coming to experience the gifts that can then arrive requires long, difficult, repeated, *practice*, and this requirement cannot be bypassed with Taylor-like false promises. All the Teachers' Convention quick and easy inquiry-in-our-classroom handouts simply induce its inevitable failure, this, in part, because of the profound power that the industrial model of empty time still holds over our imagination. "Abiding in inquiry" is therefore susceptible to becoming one more bandwagon *precisely because the ubiquity of empty time renders it thus*, demanding that it be quick and easy and efficiently implemented in a sure-fire way because, after all, "time is always running out" and we need a quick fix.

As with music or painting, or reading, or writing, or getting good at listening to others tell of their worries over those smoke spirals, or becoming deft, as a teacher, at taking care of these responses and gathering them in to our collective care and attention, these matters take tough and repeated practice to get good at. They take study and thoughtful, rigorous, scholarly work. They require sojourns that seek out those who have been in these territories before (not unlike Wabi Sabi's traverse; see Chapter 2). They take imitation and emulation and complex conversations held in the refuge of others dedicated to such work.

So arises a paradox that we cannot avoid. Our ability to welcome and care for the gifts that may come from abiding in inquiry must be cultivated "over time," and this ability must, of necessity, be cultivated over precisely the sort of time it presumes to cultivate. Unlike the immediate-yet-empty promises of procedurally-driven Taylorism ("just do this and it will work"), full-filled time is not available as a procedure to simply be obediently and mindlessly followed. It is available as a mindful practice that takes precisely full-filled time in a living field of practice (another allusion to that agricultural stubbornness) to cultivate that practice and reveal its yield. Full-filled time is thus embedded in the fullness of the very topographical fields, the "topics" being investigated and is therefore only available to be experienced and practiced through our involvement in traversing those very territories and seeking out the fullness of them. Its yields are only "won by a certain labor" (Ross & Jardine 2009) proper to those (unfragmented, unsequenced, dependently co-arising) fields. It therefore requires engagement in those topics and the time and attention and devotion *they* demand of *us*.

So part of this practice involves precisely what is at the heart of the profession of education as a whole: *study*, and, in our present case, a study of these strange histories of temporality and how they have infused our imaginations and pedagogical practices. This is difficult, detailed (albeit, in its own way, pleasurable and, so to speak, "fulfilling") work, but that is not a sign of an error in efficiency but of the worthwhileness of the object under consideration and the project of labor undertaken. Good work takes time because good things are complex and demanding of us and our attention and devotion. In this chapter, we are struggling with a topic that is and has perennially been tough in its telling: time, the worth-whileness of learning, and how that worth is to be sought and cared for.

Abiding in inquiry thus exists in the strangest sense of temporality. Full-filled time is the time *of* the work being done, the time belonging to the fullness of that work and its rich territoriality. It is migratory and returning, like gathering flocks. Or perhaps like those Golden Eagles greeted, praised, and lovingly counted and re-counted in the mountains to the West of Calgary each year, year after year. Each new count gathers up where we were "before" and adds itself to the story already underway, as do those volunteering for the count and those they tell of what they have done. The territories of traverse, the arcs of seasons and links between weather and food sources, the maps drawn and re-drawn, the missives sent north and south along the migratory routes through emails and websites full of images, counts, concerns, advice—all this happens, not precisely "at" the same ("now") time of some empty "present" but rather "in" the same, shall we say, *presence*. The odd thing is that this is a presence that is not just "the present." It is a presence that forebodes or anticipates, that remembers and recalls and gathers, and it is experienced as having long-since done so and as continuing to do so. This, therefore, is not a matter of one isolated piece of information (first before and then) after another, but of a gathering into a while of time. That work done "before" becomes part of the present, because the "present," in the life of such work, is no longer simply a "now." "Being present" has a different meaning:

> The concept of tarrying temporality [full-filled time] allows us to see how 'the present' posited as this (impossible) dimensional entity [a "now], has been

> abstracted from 'the present' as *that which is fully here for us*, as a matter about which we have, so to speak, presence of mind. This is the distinction between empty and filled time that Gadamer's [1970] title ["Concerning Empty and Fulfilled Time"] alludes to. (Ross 2006, p. 110).

When such work goes well, we lose a sense of the dominance of the empty measure of clockwork time and fall into the time of the story, of the work, of the place itself, of "that which is fully here for us." Even if the Eagles are "late," this is not the lateness of industrial assembly, but a lateness that sits in a long, migratory gather of time and a longer gather of tracing these migrations over the years. Even if the migrating Eagles are "late" in passing overhead, this lateness finds its measure and its significance only in the wider arcs of "that which is fully here for us" in such a worthwhile study—seasonality, these birds, this place and its routes for migratory passages, and the gatherings of memory and knowledge and experience, of possible reasons and anticipations, perhaps even ecological dreads, that such a place gathers unto itself and induces and protects.

Thus, key to "abiding in inquiry" is working in the *presence* of a topic and working to bring that topic to presence. It thus involves *the gatherings and the re-gatherings*, where what we've found, where we are "now," and "what is to be done next" are gathered together into the presence of "that which is [emergently] fully here for us"—the topic being investigated. These gatherings cultivate precisely this deep dependent co-arising key to worthwhile work: the sort of presence of mind that that which is fully present requires if it is to be thus present and if we are to have presence of mind about it. There is a phenomenologically undeniable sense in which each gesture of the study and exploration of a topic happens within "the same" full-filled time of "that which is fully here for us"—"presence (of mind)" not simply "the present [now]." Such presence is thus linked to forms of practice, but it is not fully adequate to think of this practice as occurring over a long string of "nows" but as itself recurring, remaining, somehow, in the same temporal locale of emerging presence.

Pushed one step further, this sense of full-filled time requires that we understand the topics entrusted to teachers and students in school as constituted, not by lifeless fragments and bits and pieces, but by living disciplines, live inheritances and fabrics into which our lives are already woven. This sense of aliveness (*lebensweltlich*) is thus key to resisting the rush of empty time: in inquiry we are dealing with "entit[ies] that exist only by always being something different. [They are] ... temporal in a radical sense. [They have their] being only in becoming and return" (Gadamer 1989, p. 123). The topics we explore with students—betrayal and fidelity, perhaps, or "democracy" and its exportation to the Middle East and further East, or the curves of highway off-ramps and their mathematics, or the political spins of statistics, or the fragility of the Weaselhead marshlands in Calgary, and on and on—are therefore no longer properly understood as simply fixed and finishing objects to be assembled or delivered, but are, rather living and often deeply contested "inheritances" that have been handed to us and to which we have be handed. They are living, ongoing gatherings into whose life of gathering we must enter into in order to come to understand them, in order for each of us to "gather" something of these gatherings. They are stories into whose tellings we must step and add ourselves. To understand, then, is to "further" (Gadamer 1989, p. xiv) the gathering—just like listening to and understanding the story of Fideles means becoming caught up in the wariness of cats, or the signals of smoke (this is but one example of how a good story has multiple ways of gathering—"only in the multifariousness of such voices does it exist" [Gadamer 1989, 284]). This requires that, in order to come to understand what is going on with a particular topic under investigation in the classroom ("Something is going on, [*im Spiele ist*], something is happening [*sich abspielt*]" [Gadamer, 1989, p. 104]), I myself must enter into this "ordering and shaping" (Gadamer 1989, p. 107), because *that* is what any living topic *is*: an ongoing, still-gathering ordering and shaping, not just a fixed and already finalized order and shape. Even the seemingly finished formulae of, say, quadratic equations, are always appearing and re-appearing in the life of the life world, demonstrating and re-demonstrating the character and limits of their applicability and usefulness.

Abiding in inquiry thus requires practice, requires engagement. Temporally put, "to be present means to *participate*" (Gadamer 1989, p. 124). Full-filled time thus links coming to know a living field of work and its gatherings to the transformation of the one coming to know into someone who "know[s] one's way around" (Gadamer 1989, p. 260): "this means that one knows one's way around *in it*" (p. 260), in the gatherings of and in the dependently co-arising gathering presence of mind regarding, a living field of work.

There is a sense, here, that time, in such gatherings, "slows down." But this isn't quite right. This isn't simply a slowing of the clock:

> It is not merely one's "taking time" to linger over something, as in the slackening or slowing down to contemplate. [This full-filled] temporality . . . is not a function of lackadaisical, meandering contemplation, least of all passive in any way, but is a function of the fullness and *intensity* of attention and engrossment (Ross 2006, p. 109).

We become enthralled and "enveloped in a time that does not pass" (Ross 2006, p. 106), a time described by Hans-Georg Gadamer with the German term *Verweilen*—translatable as "tarrying" or "whiling" or "gathering":

> In this tarrying the contrast with the merely pragmatic realms of understand becomes clear. The *Weile* [the 'while' in *Verweilen*, tarrying] has this very special temporal structure—a structure of being moved, which one nevertheless cannot describe merely as duration. In it we tarry. (Gadamer 2001, pp. 76–7).

This possibility of, shall we say, "absorption" and being moved and addressed and, shall we say, summoned or beckoned by the work itself, is phenomenologically familiar. When the work undertaken is worthwhile, the inquiry, the topic, the images, the ideas, the story:

> truly takes hold of us. [I]t is not an object that stands opposite us which we look at in hope of seeing through it to an intended conceptual meaning. Just the reverse. The work is an *Ereignis*—an event that 'appropriates us' into itself. It jolts us, it knocks us over, and sets up a world of its own, into which we are drawn, as it were. (Gadamer 2001, p. 71).

This link between worthwhile work and the while of its full-filled time has a wonderful, pedagogically rich, double movement: the thing we tarry or while over, as well as our ability to experience it " ... becomes ever richer and more diverse. The *volume* increases infinitely—and for this reason we learn from [it] how to tarry" (Gadamer 2001, pp. 76–7). In inquiry, the topic becomes richer and we who gather around it and while over it become more and more able, through participation and practice, to experience that richness and diversity. As the topic becomes richer and more diverse, we, as St. Augustine put it, become "roomier" (cited in Carruthers 2005, p. 199).

There are important pedagogical hints in all this.

First, it is not just that good stories and good work ask us to linger over them and return and gather. Such things *teach us* how to tarry or "while" by *requiring this of us*. "The 'self' is *forgotten* in the experience of tarrying" (Ross 2006, p. 112) as we become absorbed in the while of the work. This is why I've left these heated signifiers—"good stories," "good work" undefined, because they gain their signification and significance only through the concerted concern for and dedication to the question of what it is that is *worth* whiling over. "Who is to say what is worthwhile?" cannot be genuinely asked from a cynical distance by a self that is full of self-consciousness and who always wants to be first and who refuses the risk, refuses to let themselves enter the fray, the *Spiel*. "Something is going on, (*im Spiele ist*), something is happening (*sich abspielt*)" (Gadamer, 1989, p. 104)—the isolated and bereft, post-modern, cynical, self appears, now, as nothing more than a bullying, self-obsessed spoilsport who has fallen hook, line and sinker for the story of fragmentation and cool, safe, self-congratulatory distance and alienation. In short, the subjectivization of the question of worthwhileness is a hidden product of precisely the fragmentation, isolation and sense of public powerlessness and stranded-ness that codified in Taylor's efficiency movement.

Second, good work in the classroom becomes akin to an object gathering mass that, as it gathers, starts increasing its "draw" of attention. Its "gravity" increases as our knowledge and experience of it increases, which thereby draws our attention even more strongly. This, in fact, is a familiar experience. The more attention I give to this new work of art that has entered my house ("The Aviator"!), or the more attention I give to writing this chapter, or the more often I read these strange writings of Sheila Ross or Hans-Georg Gadamer or Jane Yolen, the more they attract my attention: "Having become more experienced about some thing through whiling our time away over it has a strange result: *what is experienced* 'increases in being' (Gadamer 1989, p. 40)" (Jardine 2012e, p. 190). The more often those Golden Eagles are noted, the stronger the draw of that noteworthiness, and until I myself enter into that sway, that noteworthiness is not yet "present." It must be cultivated to be experienced. It does not lie there openly and anonymously available. That is, its coming to presence asks something of me. This is why Gadamer (1989, p. 122–3) parallels this whiling time to a sort of "festive" time, where, over time and through cyclical returning, something like a "tradition" is set up and why he suggests that "belonging" in this arc of telling and re-telling is a condition of the story's draw (p. 262) ("a sense of 'nativeness,' of belonging to the place [Snyder 1980, p. 86]). We find this in classrooms where a time for storytelling—or any other experience considered by those gathering to be worth while—has been carved off and protected and reliably repeated. This is far from being simply the setting up of a classroom "routine" that is done simply for the sake of familiarity and predictability:

> [the world] compels over and over, and the better one knows it, the *more* compelling it is. This is not a matter of mastering an area of study (Gadamer 2007, p. 115).

Thus, getting to return to something worthwhile has its own attraction which then teaches us about the worth of this kind of returning. Good, worthwhile work creates a desire for good, worthwhile work and an impatience with trivial things that are not only not worthwhile but that ravage and atrophy and betray our keenness for worthwhile things. As Chris Dawson (1998, p. xxvi) notes in his "Translator's Introduction" to Hans-Georg Gadamer's *Praise of Theory*, this links to the hermeneutic interest in old Greek ideas of "the beautiful": "Any beautiful thing has a radiant elegance about it which ... points beyond itself and drives us to look for further elegant unities in other things." Again, then, whiling over a good story teaches something more than the tale being told. It "attracts the longing of love to it. [It] disposes people in its favor immediately"

(Gadamer 1989, p. 481) and it disposes us to seek out and surround ourselves with such things. It disposes us to clear out the junk we have surrounded ourselves with and the junking of our lives that such junk induces. It teaches us—teachers and student alike—not just something about this or that particular topic, but also something about the worth of whiling itself and what it requires of us and what happens when we strive to surround ourselves and fill our lives with things worthy of, quite literally, spending our lives on.

Worthwhile things are thus secretly honoring of our sense of our own finitude. Pursuing "good work" involves becoming more and more aware of the frail fabric of things, of the vulnerability of knowledge and scholarship itself in this rushing, often degrading, trivializing world and of the knife-edge limits of learning that is, as living, always and inevitably up against the edge of impermanence. Up against this edge, that abiding in inquiry *happens at all* in the real world of real schools is downright miraculous, and this makes it all the more beautiful and all the more honorable to be part of. The mindfulness of inquiry often requires bloody-mindedness and refusing to expend ourselves in the ever-accelerating rush of empty time that is deliberately designed to never be satisfied and to produce in us a cynicism about any viable alternative:

> Gadamer's uniquely concrete account of the temporality of tarrying facilitates ... a view of [human] continuity that is profoundly participatory. It gives the deadening abstraction of cultural alienation a specific meaning: participation is, so to speak, only a thought away. This corrects the view that the existence of the cultural artifact obviates or delays indefinitely any need to inquiry into human continuity oneself, as though the cultural artifact performed this custodial service for us by itself. It becomes possible to see how identifying human historical continuity, not *with* the hermeneutical event [of tarrying and gathering *for oneself* and *with others also gathered*], but *as* 'elsewhere' [already in the book, already known nor remembered by others, easily accessible if I want to on-line at a moment's notice, so why should I bother myself with learning it or tarrying over it?], merely lends itself to a self-understanding as helplessly alienated. The richness of the time of tarrying ... provides a resource for revising our terms of engagement with the question of, say, the tragedy of our post-modern 'condition' by offering a way past the prognosis that this condition cannot be corrected. (Ross 2006, p. 113).

One Ending

"The temporality of tarrying serves as a backdrop against which [empty] time now becomes suspect" (Ross 2006, p. 109).

"So that sitting there, listening like that becomes part of the story too, just as I am added when I tell it, as anyone will be, each version a journey that carries us all along" (Wallace 1987, p. 47).

Thus the great migratory conspiracies of storytelling, the telling from breath to breath (Illich 1998), each breath, of course, original, irreplaceable, and necessary in breathing life into the while of worthwhile things. "If the story does not have that breath of life, all the journeying, all the history, all the mystery, are for nought" (Yolen 1988, p. 10). Children come to school already breathing, already teeming with words and images, with loves and lives and fears, and tales of cats and water trickled through dirt. The stories of what happens next are always then created "in between," in the liminal space that is neither inside nor outside, neither mine nor yours but beckoning of us both out into the fabric of the world and its "to and fro movement" (Gadamer 1989, p. 104).

To reiterate: "*The true locus of hermeneutics is this in-between*" (Gadamer 1989, p. 295). The true locus of pedagogy is this playful, venturous in-between.

The King is already at the window.

Old Thomas is already hidden in the curtain folds, already listening. And something about cats rattles through each ancestral memory trace that we each bring to this story's gathering presence.

. . .and Another

> Thousands of years ago, the work that people did had been broken down into jobs that were the same every day, in organizations where people were interchangeable parts. All of the story had been bled out of their lives. That was how it had to be; it was how you got a productive economy. If ... employees came home at day's end with interesting stories to tell, it meant that something had gone wrong. The Powers That Be would not suffer others to be in stories of their own unless they were fake stories that had to be made up to motivate them.
>
> —from Neal Stephenson's *Anathem* (2008, p. 414).

Thus this, "... lifted from one teller's quilt and sewn into another" (Yolen 1988, p. 5).

Several years ago, in a Grade Two classroom, I asked a child "Why do you read?" His answer?

"In order to improve my reading skills."

8. *Echolocations*

Jackie Seidel, David W. Jardine, Deirdre Bailey, Holly Gray, Miranda Hector, Judson Innes, Carole Jones, Tanya Kowalchuk, Neelam Mal, Jennifer Meredith, Carli Molnar, Peter Rilstone, Trish Savill, Kari Sirup, Lesley Tait, Lisa Taylor, and Darren Vaast

The True Names of Birds

There are more ways to abandon a child
than to leave them at the mouth of the woods.
Sometimes by the time you find them, they've made up names
for all the birds and constellations, and they've broken
their reflections in the lake with sticks.

With my daughter came promises and vows
that unfolded through time like a roadmap and led me
to myself as a child, filled with wonder for my father
who could make sound from a wide blade of grass

and his breath. Here in the stillness of forest,
the sun columning before me temple-ancient,
that wonder is what I regret losing most; that wonder
and the true names of birds.

Sue Goyette (1998, p. 11)

This chapter was originally composed as an oral presentation at the *Eighth International Imagination in Education Research Conference* in Vancouver British Columbia, July 2013. We begin with Sue Goyette's lovely poem that inspired our work and was one of our texts.

In the middle of a set of four graduate courses that we undertook from May 2012 to April 2013, and in which we explored the hermeneutic and ecological roots of rich inquiry in the classroom, Neelam Mal brought in a book entitled *Nightsong* (2012) by Ari Berk and illustrator Loren Long.

Its immediate appeal was in a particular passage that seemed to echo the experiences we were now starting to repeatedly have as our explorations continued:

> The sun had set, and the shadows clinging to the walls of the cave began to wake and whisper.
>
> "Chiro? Little wing?" the bat-mother said to her child. "Tonight you must fly out into the world, and I will wait here for you."
>
> "But the night is dark, Momma ... darker than the moth's dark eyes ... darker even than the water before dawn," the little bat exclaimed, twitching his ears this way and that.
>
> "I know," whispered his mother.
>
> "And when it is that dark outside, I cannot always see," Chiro admitted, stretching his wings.
>
> "There are other ways to see," she told him, "other ways to help you make your way in the world."
>
> "How?"
>
> "Use your good sense."
>
> "What is sense?" the little bat asked.
>
> His mother folded him in her wings and whispered into his waiting ears, "Sense is the song you sing out into the world, and the song the world sings back to you. Sing, and the world will answer. That is how you'll see."
>
> ... And then she let him go. (n.p.)

We'd all started to experience something of this, of reading out a passage in our class from David G. Smith or Cynthia Chambers or videos and writings of Maxine Greene, or of Thich Nhat Hahn speaking with David Suzuki, or Edward Espe Brown speaking of Buddhism and bread-making, or instructions on how to make a mandala, or Sue Goyette's and Don Domanski's poetry, and in each gestural move, hearing these things echo off of memories of previous classes, turning into small classroom experiments that someone documented online of their Grade Two children's mandala work, or a story of mornings and families and birds in the back yard, or an anecdote or thread of family history, years ago, or meditating on Styrofoam cups, or a poem whose authorship had faded. The longer our class went on, the richer became these echoes until, near the end of that phase of our work together, vast echoing silences often overtook our gatherings as the voices of "responding and summoning" (Gadamer 1989, p. 458) gaped open and drew us in. Just as often, gales of laughter or gasps of deep recognition, and ever-increasing time spent with each other over food and the most lovely commiserations. Teachers, kids, hard work, and finding deeper and deeper soils as we went. Finding that the seeming "radicalness" of what we were pursuing actually linked back, etymologically, to a sense of "rootedness." Latin: *radix*.

Like radishes!

> From 1650 onwards, "going to the origin, essential" (On-Line Etymological Dictionary).

Jackie Seidel told us about new research being conducted on whales and their echoing songs and it became clear, once again, not that "'this is that' but this is a story about that, this is *like* that" (Clifford & Marcus 1986, p. 100). We heard this tale as a great allegory to our work with students and with fellow teachers, an allegory to our developing and newfound refuge in echo-proximity to each other. We aren't crazy, and our seeming radicalism is not against the grain of things but with a grain deeper and more sustaining than the shiny false promises of much of our profession.

These whales not only locate each other in a vast three-dimensional space, but also can locate the locating of other whales locating each other. An email from Jackie to David:

> I think the "location" part is interesting because it hints towards how much it matters that we know *where* we are and what the contours of that space are and who is sharing it with us. There is a kind of present-ness to it all that relates to what we've been talking about in our class... to be 'in' time and 'in' place requires a different kind of echoing learning/knowing. When the whales (or bats) are echolocating themselves, it is for right now, to learn their way about the place they are actually in, not some other place. Scientists who study whales say they know so little about how any of this works, or what the whales are actually experiencing. They know they can "hear" distant pods and are probably collecting information from them as well... they don't know if the entire ocean is one huge echo communication space for cetaceans. They haven't evolved at all for something like 50 million years since they went back into the water from land. It's so amazing. They have highly developed senses and social abilities, possibly greater than humans. There are so few of them left compared to 100 years ago. But whales live a very long time so there would be whales alive who probably "remember" when they were abundant. The emotion part of the whale brain is at least as developed as humans. I wonder how they think about what has happened to them?

Their echoes, it seems, can reach right into the flesh of other whales (Warren, 2012).

. . . . . .

Shut eyes
Take breath
Breathe ...
... Pause ...

Open, move
Sway,
Poised,
with grace.

Miranda Hector

. . . . . .

Something happens to you as you read the words written by others. They become intertwined with the fabric of you. They carefully wrap themselves around your neurons, heart muscles and bones of your body, ensuring their survival with you.

As I read my grandmother's words, I find a way to visit the echoes she has left behind. I become the precocious child who found herself at odds with her mother's Victorian understandings. I am she and she is I. I am the emigrant to Canada travelling by train alone, pregnant and full of both fear and excitement. I am the new mother in a new country who writes home with feelings of adventure mixed with a touch of regret. I am the old woman who has maintained her spirit and still finds herself involved in the occasional mischief. She for a brief time, is at home within me.

Her words and her being will continue to reside and echo within me.

Watching Maxine Greene on the screen created a deep wish within me. I want her to be my grandmother. Could her spirit and essence help guide me? With time and thought, I realized, she already was.

Lesley Tait

. . . . . .

Themes develop, overlap, expand and repeat
Experiences multiply and perspectives change while the soul remains undisturbed.
paraphrasing Maxine Greene (2001, p. x)

As I consider the underlined and fervently starred passages from years ago, I am reminded of a former self. A self changed by experience. Made different for the better and in many ways, worse.

The passages beckon. They call me back to a time that was simple. Idyllic. As a new teacher, I knew little and held onto those passages with the desperation that inexperience brings. They were all I had. I did not know curriculum, rubrics or competencies. I was not ruined by instructional minutes, outcomes or benchmarks. Ignorant to the science of teaching, I was free to lose myself in the art of being with children in unmeasured moments.

Still, to share what we notice, free from the strings of accountability. That was all I had. That was all I knew.

The longer I teach, the more I realize what is unimportant.

Kari Sirup

. . . . . .

"Sense is the song you sing out into the world, and the song the world sings back to you."

Ari Berk & Loren Long (2012, n.p.)

. . . . . .

My children have been waking up early lately and it has been driving my husband, Chad, and me a bit crazy, well a lot crazy. We've been taking turns getting up with them while the other stays dozing and warm for a little bit longer. This morning was my turn to stay in bed and sleep. We live in a small house, so I can hear everything being said down the hall. It went something like this:

Chad: Buddy why are you up so early?
Jake: The birds are singing to me dad.
Chad: Hmm, let's get something to eat.
Jake: Why are the birds singing to us dad?
Chad: I don't know. Lets get some breakfast.
Jake: I love when the birds sing to me dad.
Chad: What do you want to eat for breakfast?
Jake: Can we look and see if we can see the birds that are singing to us dad?

Lisa Taylor

. . . . . .

We have completely forgotten that the trivialization of joy in learning isn't the real world. In our rush for scholarship and superficial memorization of out-of-context data, we unthinkingly absorb the message that this just is the way things are. It isn't the way things are, it's the way we've made them. History has handed us over to a way of living and learning that doesn't allow us the opportunity to know things properly.

Deirdre Bailey

. . . . . .

Our topic is echolocation. The process of navigating our way in the world by sending out pulses of sound and interpreting the returning echo. If you do this by yourself, then what you get reflected back is only a measure of your own perspective alone ...

Darren Vaast

. . . . . .

> Catch only what you've thrown yourself and all is mere skill and little gain. But when you're suddenly the catcher of a ball, thrown towards you, towards your center . . .
>
> Rainer Maria Rilke, from the frontispiece of Hans-Georg Gadamer's *Truth and Method* (1989).

. . . . . .

... when you are in a community of others all on the same journey then your sonic senses are amplified, modified, and shaped by the perceptions of others around you. You get a clearer view of the world around you—a more three-dimensional understanding if you will. As teachers, we experience this every year.

Darren Vaast

. . . . . .

We read Jackie Seidel's "A Curriculum for Miracles": "With more breath. I learned to expect miracles and also to create space for miracles to happen. I learned that life itself is a miracle and that we are miracles, each of us" (see Chapter 1).

. . . . . .

Breath

. . . . . .

Jake: Can we look and see if we can see the birds that are singing to us dad?

(This was the miracle. "Expect miracles and create space for miracles").

In the end, all four of us wrapped ourselves in blankets, and sat quietly in the middle of the lawn, looking for the birds that were singing to us.

Our next step is to name them and to recognize their songs.

Lisa Taylor

. . . . . .

"... they've made up names
for all the birds ...."

Susan Goyette (1998, p. 11)

. . . . . .

Shut eyes
Take breath
Breathe ...
... Pause ...
Open, move
Sway,
Poised,
with grace.

Miranda Hector

. . . . . .

A Simple Snowstorm

It was April 11th, an ordinary day that the children called magical.

The children were working at their desks on a task that neither myself nor the students can remember now. I think it was math. Outside our classroom windows, April snow was falling steadily. Giant, delicate flakes of snow were dancing through the open sky, seeming to mock our still, rigid bodies, stifled by hard desks. It had been one of those dreaded indoor recess days, not because of cold, but because of how wet and puddly it was outside. Something about not wanting the children to get wet in the puddles.

The snow was now falling more heavily, almost blizzard-like. We quietly got our jackets on and snuck outside the back door of the school.

Together, we ran outside, breathing in the crisp air and instantly feeling the soft blizzard powder tickling our faces and covering our hair. The children opened their mouths to catch the wet flakes on their tongues. We laughed at how silly we looked covered in the snow, laid on our backs and watched the snow falling towards us, and we darted through the snowflakes, chasing one another.

We became soaked.

"It's like we're having a shower with heaven water!" one child remarked joyfully.

"That was the best day of my life!" one boy announced when we arrived back in our classroom.

Carli Molnar

. . . . . .

> Let's consider an ecological analogy to these isolated, deliberately unsurrounded, worldless, school activities. Consider this Styrofoam cup I'm just about to throw away. Any relations of it or to it cannot be cultivated, chosen, cared for, remembered, enjoyed, either by us or anything else that surrounds it. I cannot easily become composed around such a thing and it does not ask this of me [consider Lesley Tait's favourite passage that echoed in our class again and again, from a text composed by Tsong-kha-pa in 1406 in Tibet: "I compose this in order to condition my own mind" (2000, p. 111)]. There will be no mourning at its loss or destruction. It is not something to be saved or savoured. It does not show its having-arrived-here and we have no need to try to remember such an arrival. All trace of relations and endurance are seemingly gone. It appears, and then disappears, and its appearance is geared to not being noticed.
>
> This Styrofoam cup does not *endure*. It does not age and become becoming in such aging. It breaks.
>
> It cannot learn from its surroundings and show the wear of such learning on its aging face. And therefore, in its presence, the prospect of pedagogy is turned away.
>
> In fact, it is produced *deliberately* in order to *not* last, *not* hold attention, *not* take on character, *not* arouse any sense or possibility of care or concern. *It is deliberately produced in order to not be remembered.*
>
> It is deliberately produced of forgetting. It is *Lethe*. It is lethal.
>
> —*David W. Jardine (2012c, p. 139), from "Figures in Hell."*

. . . . . .

One day, Trish asked: "What is the Styrofoam cup in the classroom?" We all stopped still.

A Styrofoam cup is disposable, yet permanent, like the worksheets and the fragmented activities so many children do in classrooms. The worksheets, the spelling tests, the unimportant tasks are disposable, yet the effect they have on children is often permanent. As the Styrofoam cups stack up, our ability to imagine, invent and create dwindles and the world as black and white becomes reality.

We can't care for Styrofoam cups and children can't care for disposable work.

Carli Molnar

. . . . . .

When I find a crossword puzzle or a Sudoku that has been partially filled in, I am left with an empty feeling inside. The puzzle is ruined. It's fouled.

It is a single use, a one-time thing and when others have been there first, it ruins it for me. The mystery of self-discovery is gone. There is no more life in it anymore.

They are over even before you begin.

Darren Vaast

. . . . . .

Styrofoam Cup
Conference rooms to birthday parties
Hot tea, lukewarm juice
So small, never enough.
Stark white, light and airy
Statically charged,
Buoyantly held.
Sipping, nibbling, chewing
Squeaky, hollow
Broken pieces
Fingernails scratching, imprinting.
Imprinting the cup
Imprinting the planet
Biodegradable? No, degraded.
Chemicals leeching into us
Cancer

Holly Gray

. . . . . .

It "cannot weather or age. [Its] existence is hurried by the push of obsolescence as one generation succeeds the next within a few months. [Its] suffering is written on [its] face" (Hillman 2006, p. 39). This is no longer the gentle, heart-breaking suffering of being left in peace to impermanently become (de)composed in the embrace of the world. It is a deeper suffering produced of illusion, one that conceals this difficult, convivial peace that comes with the experience of impermanence. Hillman (p. 39) says that this image of an object that has "no way back to the Gods" is precisely an image of a "figure in Hell." Ecologically and pedagogically understood, an image of a hellish figure cut off from what ought to be there for it *to be* its convivial, earthly self.

> This is something worse than mere illness.
> It is evil.
> "Evil is the absence of what ought to be there." (Hardon 1985, p. 136)
>
> *—David W. Jardine (2012c, pp. 143–4), from "Figures in Hell."*

. . . . . .

Cynthia Chambers, "Spelling and Other Illiteracies" (2012, p. 188):

> "Divorced from knowledge and tradition and imagination, literacy becomes a technology instead of a practiced craft ... "

Education is such a busy place. There is allure to the notion that a shiny, packaged program with laminated colour-coded flash cards could maybe help ... maybe take something off of our VERY full plates. I see colleagues reach for these things in faith that if the school, the board, the system bought it, it must have value, it must nourish our children.

Assignments are uncomplicated, assessment is easier, answers are clearer. But it's like candy ... so sweet and easy and then suddenly your teeth are full of holes.

Neelam Mal

. . . . . .

Throwaway stuff that never goes away.

Jennifer Meredith

. . . . . .

This Styrofoam cup makes me awkward and uncomfortable because now that I've touched it—I am connected to its permanent presence in this world. A presence that cannot be undone and will not fit with what is. My complaisance is a crime which is knowledge's fault because now that I know about Styrofoam I have this responsibility to live up to.

Or:

A purpose?

How many of us bustle through purposelessness, lamenting that reason for existence has been lost. What a blessing to have found such an evident purpose in this cup. Be uncomfortable. Conscious. Careful.

Learn, react, repeat.

Deirdre Bailey

. . . . . .

Lately, I've been spending time with the cup. Writing report cards, purchasing dollar store items, debating on what brand handbag to buy next, planning for Chemistry, dining with "friends".

I sit with friends over wine and dinner talking about life. About downtown Calgary. About their new Lexus. About their upcoming trip to Hawaii.

About the unethical deal that in the end turned out "ok" and made him thousands.

Styrofoam.

Lisa Taylor

......

Eternal Death

Pure white. Pristine smoothness. Seemingly innocent. Cloned sameness.

Collective indifference. Impermanent permanence. Wasteful convenience. Eternal Death.

Darren Vaast

......

One of our reliable companions in this echolocation venture was the Canadian poet Don Domanski. He told us of:

> ... what the Japanese refer to as "the slender sadness," that runs through every moment of existence, about the fleetingness of lives lived in a world where nothing can be saved. [It is] entering that state of being with a joy and wonder that comes from that very impermanency, from the absolute dispossession of everything we love and cherish. The wonder is that anything at all exists. The joy is that it does, even if it is as momentary as a human life. We can live this as a mode of attention, we can live within its movements, its cycles and treasure the phases, the round of it. (2002, p. 246)

> At poetry's centre there's the silence of a world turning. This is also found at the centre of a stone or the axis of a tree. To my way of thinking, that silence is the main importance. Out if it comes the manifestations, all the beings we call words. (Domanski, 2002, p. 245).

One of our "required textbooks" was Don Domanski's beautiful *All our Wonder Unavenged* (2007).

As our time together increased, so, too, did the silences, where the smallest of gestures or the slightest of words arrived so full as to need larger and larger spaces of silence to work out its echoing returns.

And then at the end of such a silence, someone would say, "So *that's* why David G. Smith used that phrase!" and their words would arrive as a benediction for all we had been through, each one of us hearing and feeling and understanding this "*that's* why ..." as a hint of some intimate, unutterable secret that held us together.

David W. Jardine

. . . . . .

A Pocket of Darkness

The comfort of timely darkness is banished
by artificial, unearthly glow.
On the extremes of the city the eternal, held off, awaits.
In the coulee a pocket of darkness.
Marbled pairs of reflected light,
briefly glow, then shimmer and fade out.
Alone now, they wind through tangles, relentless,
and re-emerge into one.
Call up to the creators; we are here, we are here.
What of those who occupy the spaces in between,
and linger along the precipice.
Outlines given vague shape by the fleeting;
ephemeral and haunting.
Await the familiar, an echo emerging from below.

Judson Innes

. . . . . .

Part way through our last class together, we sat listening to the laughter and the lovely blurs of conversations after coming back to the classroom from our break for food.

"Listen. They *love* each other."

Jackie Seidel & *David W. Jardine*

. . . . .

Shut eyes
Take breath
Breathe …
… Pause …
Open, move
Sway,
Poised,
with grace.

Miranda Hector

. . . . .

Big flakes of snow meander from the gray sky on this spectacular winter day, the perfect time. We walk slowly through the deep snow. This grade one group assure me that this is the best way to uncover the secrets of their new classroom.

They are startled when birds fly so close we can feel the air beneath their wings. The sun peeks through the grayness, radiating bright streams that tickle our faces with warmth and make the snow sparkle like gems.

We stop to feel its power. Our quiet conversation is interrupted by strange movement in the bushes.

We are serenaded by the music of the ducks.

... to be part of this peaceful classroom.

They have interrupted the rush.

Trish Savill

. . . . . .

After all the brightwhite thoughts of disposable Styrofoam cups that deflect and reject our love and care, objects that cannot age with our own aging, objects that provide no surroundings, no housing of memory ... Jackie Seidel brought in Don Domanski's *All Our Wonder Unavenged* (2010) and we read, aloud, several times, his "Disposing of a Broken Clock." He has graciously granted us permission to cite this wondrous poem:

. . . . . .

Disposing of a Broken Clock

I put two final drops of oil in the mechanism
which is placing a word on either side of time
I anoint it twice bless it in a proper manner

to pay homage to those cogs and gears
that held the hours in a drop of oil for so long
like dark wine that settled on an altarpiece
of moveable stirrups
coiled springs and mandibles

I carefully removed the screw that controlled
the weightlessness of the future and the one
that counterchecked the heaviness of the past

I wipe its face clean of hands and numbers
so it looks like the moon sighted through
longing what the eyes have to endure

I remove the mainspring which shivers once
in my palm and then is still the stillness of harm
done harm's way straining to be silent

finally at its centre I build a small campfire
to warm the ones who will come much later
those migrants those small beasts who circle
us endlessly who follow the ticking of the grass
and the straw that overtakes the wind
who know only that an essential time lives on after us
who bow to the timepieces lowered onto their hearts
the continuum of water and the laze of stones.

Don Domanski (2010, p. 67)

. . . . . .

Peter Rilstone helped us linger over the images of anointing with oil, of penances and patience over the dead body, clean and clarified and let go into its proper disposition. Peter has recently retired from the local school board, but not from teaching. He provided exactly what our class needed—the great lesson that now, retired from the day to day work of schooling, the best thing to do was to learn Latin.

We ate up Domanski's words again and again. Something cared for that is properly let go back outwards into all its relations. All the slow words of giving up this thing in the world that had lasted, that had gathered memory and care and love around its face.

Over time gathered and not rushing.

There is one image in particular in "Disposing of a Broken Clock," of wiping the face of the clock clean of hands and numbers that sent an unexpected shock through the class, an image of a young child come up close to have their face wiped cool and fresh and clean in summer heat, but then the child, too, who has died whose face is wiped and anointed for burial, with all our thoughts arced back to Jackie's "Curriculum for Miracles."

So Domanski's old and broken clock is properly taken care of, properly ushered back to the gods that made it, back to "the continuum of water and the laze of stones."

We are miracles, each of us.

David W. Jardine & Jackie Seidel

. . . . . .

For a thousand years scholars have esteemed the Biblical commentaries of the theologian Rabbi Rashi (1040–1145). His work amalgamated his approach to interpreting Jewish scriptures with one of his teaching practices. Of the former he reflected that for every Biblical passage there are a dozen interpretations. Of the latter, he usually responded to a student's query by posing another question. And thereby he produced his timeless work which was a compendium of his students' and his own reflections.

During our course "Ecological Consciousness and Inquiry in the Classroom," Jackie and David introduced our class to the poetry of Don Domanski. On the first reading of each of Don Domanski's poems, his writing made no sense to my classmates and me. Troubled by the apparent incomprehensibility of the assigned readings, when we requested a key to interpreting the poetry, our teachers sent us back to the drawing board with questions such as Rashi might have asked, "What does it say to you? What do you think it means?"

While a colleague and I drove to and from classes, we exchanged our interpretations of Domanski's poems. Invariably we shared common thoughts, but each of us also had different insights, based on our unique life experiences and bodies of knowledge. The mingling of our differences opened entirely new inroads to appreciate the richness of what we had read, which led to further understanding of ourselves and our human condition. It is remarkable that the car survived our inspirited dialogues (distracted driving!) during the weekly trips between home and class.

My first readings of Domanski's "Disposing of a Broken Clock," in which he eulogized a cherished timepiece, struck me as the account of a rather senseless act:

> "I put two final drops of oil in the mechanism ...
> I anoint it twice bless it in a proper manner
> to pay homage to those cogs and gears
> that held the hours in a drop of oil for so long
> like dark wine that settled on an altarpiece ..."

This poem reminded me of former possessions which I sorely missed when they were broken, lost or cast away, such as a medal, which my grandfather had given to me, and mysteriously disappeared when I was in Grade Four; and in adulthood, each of my Mazdas, in which I sat, before trading them in, and recalled memories of passengers, some no longer alive, with whom I had shared trips near and far. But never had I 'wasted' oil upon an object that I was about to discard. I puzzled, "Why did Domanski lubricate his irreparable clock's mechanism before disposing of it?" Like many of his images, their seeming incomprehensibility during the first readings struck me, and did not let go.

But sometimes time transforms what seems immediately illogical by striking chords of deep realities below and within. Domanski's words, 'anoint', 'bless', 'homage', 'wine', 'altarpiece' seemed to set apart and elevate to a sacred place his worn out, no longer functional clock. These words evoked a connection between his 'wasting' oil and a similar accusation leveled against a woman who poured precious nard upon Jesus before he died (Mark 14:3). And "*christus*" means "one who has been anointed"—that word burst a floodgate of thoughts and questions, sent me to books and Google, roused ripples of connotations, connected heretofore disparate bits of knowledge and ideas, and discharged a surge of adrenaline. Ah, the delights and highs of learning, and sharing insights with others!

Domanski's words, 'anoint' and 'oiled' seem inextricably linked. Traditionally, priests have anointed the sick before they die. Bishops have anointed monarchs with oil. The common word for oil in Latin is *oleum*—that which we pour on salad. But Latin has a second term, *unguentum*, from which comes our word 'unguent', a word that beautifully, even onomatopoeically, falls upon the ear. *Unguentum* refers to a precious oil, and also means salve, ointment, perfume. Moses anointed the first Jewish high priest with precious *unguentum* (Exodus 30:22–32). Psalm 133 reflects about this event: "Behold, how good and joyful a thing it is, wherever people dwell together in unity! It is like precious ointment (*unguentum*) upon the head, that ran down unto the beard, even unto Aaron's beard .... For there God gives His Blessing, and life for ever." The unguent or salve used for anointing or making sacred a monarch and persons on the threshold of death has a psychic and spiritual, if not physical, property of healing. Christianity has traditionally taught that the baptismal anointing of carnal flesh with water heals body and mind of their hereditary fate of falling short of a human's potential and ideal.

Through repeated reading and reflecting upon 'Disposing of a Broken Clock' my first reaction that Domanski had wasted oil metamorphosed into admiration for a symbolic act of ecological stewardship. This poet is one who does not unconsciously discard the shabbily old for the fashionable fad.

Peter Rilstone

. . . . . .

My thoughts turn to moments in time that could use a little bit of oiling. When time sticks and grinds, creaks and chugs, a little metal pot of oil at hand might smooth the gears. Not to move time more quickly. Time marches on quickly enough. Metal pots of oil with which to oil time; to ease a difficult time of waiting, to bring peace to a time of worry, to soothe a person's angst over the passage of another year—a little drop of slick oil to lubricate their troubles.

When the passing of time causes pain or sorrow for someone a wee drop of oil may soften their furrowed brow, ease their heart, slow their racing pulse.

Be where you are. Be mindful of giving time its proper due. Oil the passage of time to "countercheck the heaviness of the past" (Domanski 2010, p. 67).

Edited May 6, 2013
Warm sunny evening after a hot blue-sky day
Carole Jones

. . . . . .

Now Domanski's reverence for his clock begs questions: did those drops of oil express merely a marvel for the craftsmanship and aesthetics of an object? An awe of human ingenuity and invention? An apostrophic homage and nostalgic gratitude for a timepiece whose reliability he had trusted for ordering his life for decades? A gesture of healing or thankful dedication of an object before relinquishing it, in all its parts, to whence it came? An animistic empathy and unity (to use the descriptor in Psalm 133) with a treasured inanimate object? Or his oneness with not just the clock, but with our entire universe? A memorial of others, some may be no longer alive, whose times this clock had similarly measured? A nostalgic tribute to significant moments and events in his own and others' lives? An awed honouring of time itself? A wonder at the mystery of our vast, incomprehensibly complex universe beyond the small measure of order and sense and control that we humans presume to instill upon it? And so forth.

Beyond our interpretations of the readings in our course, a greater significance lies in how we were able, together, to approach and engage with the texts. How do we teachers mediate our charges' encounters with what at first may seem to them counterintuitive, incomprehensible, even impossible challenges? How do we prepare and open ourselves to make opportunities and spaces in which students share their life experiences, knowledge and concepts, and in which we learn from our students, even about topics that we have taught many times? How can we take up the mandated curricula in ways that honour and respond most effectively to what our students ask and articulate?

In harmony with Rashi a millennium ago, Northrop Frye explained that whenever he answered a student's question, he posed another in its wake, because queries raise us up, whereas statements lower. In this class, with this poem, and now with this chapter, we experienced ways which are as ancient as the earliest conversations among elders and young.

Peter Rilstone

. . . . . .

We don't become experienced through the application of a method because a method, properly taken up, must be taken up as if I could be anyone. [But] none of us is everyone. None of us is cut out for living just any life, and none of us will live forever. We are not perspectiveless ... I am defined but what I can thus remember, what necessarily exclusive and incomplete host of voices haunts my inner life and work and therefore haunts the world that is open in front of me. This composed and cultivated memory constitutes my openness to what comes to meet me from the world.

Deidre Bailey paraphrasing *David W. Jardine*, to his great surprise and delight (and inside joke)

. . . . . .

This is another gentle reminder that slowing down doesn't mean losing time. When else do you hear the faint, papery rattle of a dragonfly? Or smell the soil under your feet as you walk? Let us be mindful of this truth.

Neelam Mal

. . . . . .

We watched clips of a documentary film of Zen chef Edward Espe-Brown called "How to Cook Your Life" (2007):

"It's not so simple, to do what you're doing," he tells us.

We mulled over how what your doing, done well, cooks you. Careful and loving work composes me as much as I compose it (remember Lesley Tait's favourite passage from Tsong-kha-pa [2000, p. 111]: "I compose this in order to condition my own mind.")

And how we watched little video, together in class, called "How to Grow a Mandala" (Rodé, 2012), quietly drawing in our own journals. And how Jennifer Meredith and Carli Molnar did this with their own students (Grades Six and Two respectively).

And how Darren Vaast and David Jardine both got "cooked" in their own ways by this activity, frustrated with it being imperfect or rendering it asymmetrical on purpose.

"It's not so simple, to do what you're doing."

When you do, it cooks you.

"Towards you, towards your center."

. . . . . .

I promise to cut the carrot,
When I'm cutting the carrot.

Carli Molnar

. . . . . .

When I was a child I thought that maybe I'd be a teacher. It seemed fun. They had the answers to the workbook and occasionally took us on field trips.

My mom was a teacher, I liked going with her up to the big high school, to the photocopy room, being in her classroom, writing on the chalkboard, pretending I was a teacher, too.

"To read a place means you are able to dwell within it, to inhabit it, to gather from it the knowledge that makes life there possible, as well as intelligible and meaningful." (Chambers 2012, p. 187).

I believe that this is what my mom did for her students. They didn't all appreciate what it was she was doing, but many of them did. She still sometimes runs into some of them at the post office back home. They share stories with her about their work, their families, and their lives.

The original intention of my writing was not to talk about my mom, but here I am with tears in my eyes writing about her.

Jennifer Meredith

. . . . . .

Time

Tender parting and elegant flow,
a silent passing above the forest floor.
Wings raised up to gain favour,
and soften the looming perch.
Triggered now, firm thrust and smooth unfolding,
fell swoop and talon strike.
A mere moment swept away, devoured.
Echoed perhaps in a ripple of air,
or brief shrug of robust bough.

Judson Innes

. . . . . .

Shut eyes
Take breath
Breathe ...
... Pause ...

Open, move
Sway,
Poised,
with grace.

Miranda Hector

. . . . . .

the disposability of time
of time misused

letting it go as the river streams
it sits on my shoulders
heavy and waiting

an expectation of more, of wanting
calling to me to give it a name

the greeting of the two sides
    meeting in happy remembrance and sorrow
    of the expectation unfulfilled

letting it go, staying there for a while
awareness of more and of less
as shoulders sag

the open breath
sending away all that sits in opposite grain
recognition of what true names may join me.

Lesley Tait

. . . . . .

"... and then she let him go."

Ari Berk, 2012

9. *Hymn to the North Atlantic Right Whale*

JACKIE SEIDEL

"Finitum capax infiniti. Only with them do I matter." (Keller, 2000, p. 92).

eubalaena glacialis
so nearly alone now in this
in this
breath

exhale
spirit birthing deep forsaken

is it that we forgot to greet you?

still,
ocean moon gravity sloshing this giant
this giant
this giant tide in and out of the bay
again forever
are where are you there you are?
eubalaena glacialis.
chill flows enfolding
smooth bodies always movement dive and dive
and breathe   and breathe   and breathe

could there be joy in this leaping, your water-slap tails
raised to the skies? singing and singing the oceans together
these spiraling stars one becoming
listen! your voices are the motion of life

is this (y)our tears salting the waters?
this huge absence a black hole longing for the lost the lost the lost
this ship strike this fishing net poison swallowing entanglements
this radioactive sewage smeared too much suffering life
eubalaena glacialis
we are yours are you ours are you there?
related mammalian warm blooded breast feeding
oxygen breathing lifting your babies to the air.
you! with your backbone. phylum chordata. kingdom anamalia.
time's tugging knots binding this lineage through breath
and breasts of our mothers
our mothers, our mothers
remember do we remember you remember us?
did you imagine, when you gave up your limbs, when
you climbed back into the sea, that it would come
to this?

life is like this.
let us hope
it is not goodbye. not yet
eubalaena glacialis
oh sorrow
only with you do we matter

10. *Inquiry in Black and White: An Appreciation*

David W. Jardine

> When you're looking at a black and white picture ... you're looking at a graphic shape rather than the colour value—and in that sense, the image becomes stronger.
>
> —Herman Leonard (cited in Waring 2011, p. 120)

It has been a ongoing refuge of my own work over more than a decade to have small and multifarious relationships to Glendale School in Calgary, Alberta—to its teachers, its students, and its administrators, and to the work to which it has been dedicated. I have supervised nearly a hundred student teachers in its classrooms and each of them learned in their own way some of the tough lessons of trying to do "real work" (Snyder 1980) in the real world of teaching and learning. As you can imagine, this has been a long, generative and regenerative procession of ever-new faces, new ideas, new dilemmas in the troubled surroundings of schooling. It has also been a site of new, fresh solutions that hold at bay the always-impeding, always ever-so-urgent pressures of our profession and keep open the prospects of real learning and real teaching. It has been a site, too and of course, of failed experiments that always result in shaking off the tremors, learning the hard lessons, gathering up again, and venturing anew.

True to its etymological roots, there has also been constancy in this place: standing firm, faithfulness, dependability. I don't hesitate using this language, because the school's chosen motto is *Radices et Pennae*, Roots and Wings. When I think back over the genuine community that this school has become and has remained, I'm reminded of the words of Wendell Berry from his essay "People, Land and Community" (2002, p. 189), a community well-named by the title of Berry's book where this essay is housed—*The Art of the Common Place*:

> The community is an order of memories preserved consciously in instructions, songs, and stories, and both consciously and unconsciously in *ways*. A healthy culture holds preserving knowledge *in place* for a *long* time. That is, the essential wisdom accumulates in the community much as fertility builds in the soil. In both death becomes potentiality. (Berry 2002, p. 189).

Although it is a bit jarring on the face of it, I left that last line in there on purpose. It captures something of the fragility of our work as educators, because we are always facing the arrival of the new, but also the coming and going of students, of colleagues and friends in and out of a community of perennial work. Our work, when it goes well and if we dare admit it, is a beautiful sign of our mortality, but our mortality is the source of our energy and hope. The knowledge entrusted to teachers and students in schools is full of *living* disciplines, *living* ancestries, and *living* fields of venture and possibility that are ripe to enter and take up. Living worlds of knowledge need always to be remade, re-questioned, re-experienced, taken up anew in concert and solidarity with the young. This is one of the secrets hidden in rubrics of inquiry in the classroom (see, for example, the rubric for inquiry developed by the Galileo Educational Network [www.galileo.org] and used frequently by the teachers at Glendale School, as well as some of the work I have done with the founders of the Network [e.g. Jardine, Clifford & Friesen, 2006, 2008; Friesen & Jardine 2009, 2010, 2011; Friesen, Jardine, & Gladstone 2010]; Jardine 2012).

I recall, ages ago now, having a brief talk with the then-Grade One teacher at Glendale School about what I'd learned about monsters: from the Latin *monere*, to heed, to warn, to teach. Her immediate response was to ask me to come into the class and talk about this with the children. We had a lovely time talking about monsters and how they appear in books they had been reading, in dreams and in movies and in life. Strong, big scary, green, gigantic, my sister, run! I then wrote down this Latin term, and we talked and speculated a bit about why monsters appear, whom they appear to and when. What is it that they are trying to do, to say? Not *just* to frighten, but also to get our attention, to *teach*—at which point the sweet laughs started as they looked to their teacher, the monster. I urged the children to write down the Latin term in their notebooks and let their parents know that they are learning Latin at school. We all laughed, but they did take note.

Over the next few weeks, students from that class would come up to me in the school hallways with clutched books, wanting to talk about some monster that had appeared in its pages and why and wherefore. And then, unannounced and unexpectedly, four years later, a familiar face, a then-Grade Five student from the school stopped me in the hall with a then-new copy of some Harry Potter book:

"I want to show you something!"

This is the sort of work I've had the pleasure to be a small part of at Glendale School and that teachers, students, student-teachers, administrators and, yes, even some aging University professors, have come to hope for now, nearing the end of August, with the turn towards a new school year ripe in the air. *This* is the sort of community of memory and ways that supports and underwrites a lovely paper written by Jennifer Meredith (*née* Grimm) and Darren Vaast (see also Chapter 8) about some roots and wings at Glendale School (Grimm & Vaast 2011).

Look! Come with me! I have something beautiful to show you, real and living possibilities of the world around which we can shape our lives. As goes good inquiry, we'll stop there long enough to begin to experience what the place asks of us, what work, what sort of "continuity of attention and devotion" (Berry 1986, p. 32), is needed for this common place to begin to yield its secrets. Not just any place will do. Not just any work is proper.

And so, look, come with me. I have something beautiful to show you. And it just might have the real pedagogical and aesthetic import that Hans-Georg Gadamer (2007a, p. 131) spotted regarding what happens when we run into something real and substantial in the world that we desire to explore and authentically know: "You must change your life."

Yousuf Karsh (1908–2002), a Canadian photographer who emigrated from his native Armenia to Syria at age 14 and then, two years later, to Sherbrooke, Quebec. His work has been a vague, often distant companion of mine for more than fifty years. I'm always startled still by how deeply familiar are his images of Winston Churchill and Pierre Elliot Trudeau, and how something of his two images of John Kennedy got burned into my memory as a 13 year old in the newspapers and magazines that rolled out right after November 22nd 1963 (the same day, thank God, that *Beatlemania With the Beatles* was released in Canada). And, yes, black and white on purpose, and all the tangled and luscious film noir memories to boot. Black and white, shadow and light, in echo of Astrid Kirchherr's Beatles L.P. cover photo, glowing if properly wrought, with character and great purpose. Here, in this classroom, is Social Studies attended to with not only love and devotion, but also with rigorous and scholarly intent bent on the cultivation of memory. Full of echoes of an old experience nearly lost to memory that Ivan Illich traces (1993, p. 17) in the work of Hugh of St. Victor (c. 1096–1141), that if a body of work like that of Yousuf Karsh is treated properly and well, it is not only that our work shapes and illuminates it, but it, too, begins to glow and we, too, begin to become illuminated and cast into shape by *its* light and shadow. *It* illuminates *us* and makes us different that we were. Treated properly, Karsh's work bathes us in its light and shapes our lives accordingly. *This* is the deep rigorousness and discipline of classroom inquiry. This is its path, to bring the topic under consideration into its own luminousness so that our lives might illuminated and we might become better educated in the ways of the world, more sensitive to its surroundings and what is at work in them and thus in us.

Enough of these reveries. Their purpose is simply to hint that, with this seeming new arrival of "inquiry" into the educational imagination, we are not dealing with some new-fangled fad here, but an old and buoyant experience, a common place of work that goes deep into our collective lives and multiple, interweaving ancestries. We move, then, in an orbit of social study in its most strong, vivid and lasting sense, a community built of great fertility that needs our care and our dedicated and proper work if it is to remain so.

There's no need for me, in this appreciation, to reiterate more details about what happened in this Grade Five/Six class except to point to an inevitability that comes with the art of writing about such matters. It is impossible, no matter how vigilant the writing, to re-weave all the *actual* threads that made up the fabric of this or any other work. If you weren't "there," the writing is always too much and too little, always arriving a bit too late, not because of a failing of the authors, but because of the nature of experience, memory, cultivation, and the limits of words themselves. Once you see what it resulted in (check, too, the websites the authors provide), you will perhaps feel a bit like me about its near-miraculousness: that it happened seems nearly impossible in a real school under the real, embattled and difficult circumstances of our lives as educators.

And yet there it is, made up of small steps that always and of necessity fall from memory, or get remembered differently or become differently significant as the work proceeds. This is our lot and our inescapable lament, that we always wish you had been there. This, in fact, is the great free space that constitutes Social Studies itself, to bring us here and there, across geographies and histories, across lives nearby and far, and to catch teachers and students alike in the witness of a complex, abundant, living world.

This lovely work by Jennifer Grimm, Darren Vaast and their students also puts lie to those who are caught in the grips of the panic about standardized examinations, because one of its roots, one of its wings, is that no student has ever become stupider by doing good work. In fact, as Sharon Friesen (2010) has well demonstrated, marks on standardized Provincial examinations go *up* when one stops "teaching to the test" and instead teaches to the real work that *the topics* of the test require if they are to be treated rigorously, fully, and properly.

It's black and white, finally, that this red herring regarding standardized tests, and all the exhaustion that our panic over them causes, needs to be put to rest once and for all. The panic over these exams is still understandable, but it is becoming less and less forgivable. That sounds a bit harsh, but the greatest news of all is that inquiry is pleasurable, invigorating, and enlivening of teachers and students alike. And it is enlivening of those very worlds of knowledge we've been entrusted with. Admittedly, it is also *hard work*, sometimes *very hard work*, but teachers work hard *anyway*. The good news, for me, is that the hard work inquiry requires, allows teachers and students alike to delve deeply into the wonders of the world and do work whose rigorousness and thoughtfulness is, as is so clear here in this Karsh project, publicly recognizable and authentic and, in the end, joyful.

I want to end by citing an extended passage by Hans-Georg Gadamer (1986, p. 59) that I introduced at a recent professional development session at Glendale School. I cite this, first, because this was part of a speech he gave at the age of 86 in Heidelberg, Germany, and I have found it to be both heartening to and wonderfully humiliating of my own sometimes-overwhelming exhaustions and frustrations about my chosen profession. My second reason is that, as with all good Social Studies, this citation, as you'll see, is haunting my own writing and my own thinking through of what my next ventures need to be in my chosen profession.

The final reason for citing it is this: I want dedicate whatever encouragement this passage might provide to Jennifer Grimm and Darren Vaast, to the teachers, students and administrators of Glendale School, and to all those teachers and students who are struggling to find free spaces and forge solidarities in what seems like an always difficult time for the real work of becoming educated in the ways of the world:

> We should have no illusion. Bureaucratized teaching and learning systems dominate the scene, but nevertheless it is everyone's task to find his free space. The task of our human life in general is to find free spaces and learn to move therein. In research this means finding the question, the genuine question. You all know that as a beginner one comes to find everything questionable, for that is the privilege of youth to seek everywhere the novel and new possibilities. One then learns slowly how a large amount must be excluded in order to finally arrive at the point where one finds the truly open questions and therefore the possibilities that exist. Perhaps the most noble side of the enduring independent position of the university—in political and social life—is that we with the youth and they with us learn to discover the possibilities and thereby possible ways of shaping our lives. There is this chain of generations which pass through an institution, like the university, in which teachers and students meet and lose one another. Students become teachers and from the activity of the teachers grows a new teaching, a living universe, which is certainly more than something known, more than something learnable, but a place where something happens to us. I think this small academic universe still remains one of the few precursors of the grand

> universe of humanity, of all human beings, who must learn to create with one another new solidarities. (Gadamer 1986, p. 59)

Coda

After having completed what was to be the final draft of this appreciation, I was leafing through an old issue of *Mojo Music Magazine*™ and happened upon a book review that I had read before (Waring 2011). The book under review was *Jazz*, a collection of the photographs of Herman Leonard (2010). Included with the review is a stunning picture of Charlie Parker surrounded by other players cast in shadows. This and other of Leonard's photographs of Miles Davis, Billie Holiday, Louis Armstrong, Dexter Gordon and my great love, Duke Ellington, have been catching my eye for years. It wasn't until I first read this review several months ago that the obvious dawned on me, something I knew but didn't really know, that this array of work was one man's signature.

So I'm re-reading the review recently, and I read this: "Leonard ... [had] been apprenticed as a young man to legendary portrait photographer Yousuf Karsh" (Waring 2011, p. 120).

I'm not sure I can properly describe the experience of then looking back up at the accompanying black and white, shadow and light, photograph of Charlie Parker. Recognition and a stunning and vertiginous experience of familiarity:

> We do not understand what recognition is in its profoundest nature if we only regard it as knowing something again that we know already—i.e., what is familiar is recognized again. The joy of recognition is rather the joy of knowing *more* than is already familiar. In recognition what we know emerges, as if illuminated, from all the contingent and variable circumstances that condition it. (Gadamer 1989, p. 114).

This joy is part of the lovely experience of inquiry and the warrant of its hard work. Inquiry makes us vulnerable and open to experiencing what is going on around us and to how our experience gathers up in the world. It helps us *become who we are*, alert beings, ready to venture in the world.

Herman Leonard. Check out his work. He's proof that Yousuf Karsh was also a good mentor to his apprentice. It will make you want to go back and look at these Glendale portraits all over again. Incidentally, Herman Leonard died on August 16th 2010. "He had been living in Los Angeles since Hurricane Katrina struck New Orleans in 2005, flooding his home and destroying thousands of prints" (Thursby 2010).

That, of course, is another story, even though it is not.

11. The Paperwhite's Lesson Plan

JACKIE SEIDEL

Part 1—A Confession

Cold snowy day in early January.

I bring four paperwhite (*narcissus*) bulbs into our pre-service teacher education literacy class. Around the circle, hand-to-hand, crinkling brown dry skin invokes language while fingers point to tiny white shoots already emerging, whispering a promise of a plant to (be)come.

I *did* have a lesson plan.

Cleverly hoping to provoke a conversation about how a growing bulb is a narrative, and thus, how life itself is a story, my intention was to explore the ways these bulbs with their creative, generative and natural power, would stand as an example of a gift to us in our classrooms with children. I wanted to explore the ways that they would give us all the language we need, how we didn't need to be 'creative' or make 'fun' activities or worksheets, because these bulbs would themselves engage us deeply and creatively. Drawing something forth from us. From life.

***

The next week I carry the vase to our seminar, bright green shoots already six inches high, and two inches of roots pressed stretching against the glass below the bulbs. Shocked delight as I walk through the door: "Are those our bulbs from last week!?"

Despite the chatting and whispering as the bulbs were passed, there is some puzzlement. Some students are silent. As they describe their experiences of holding the bulbs, several share that they have never seen or touched a bulb before, except as food. Onions and garlic, they say. Our conversation doesn't seem to go very far or perhaps anywhere at all. The lesson plan's not working out. We explore some picture books. Later that day I plant the bulbs in rocks and soil in a glass vase, water them, and put them in the window of my office.

The vase gets passed and passed again. Hand to hand to hand.

Another week passes.

Through the classroom door I carry the vase, the plants now 15 inches high. Gloriously flowering fragrant white blossoms. I place the vase on the table.

Someone gasps.

Are those OUR bulbs? The ones from 2 weeks ago?

And then laughter.

How did that happen?! I had no idea! I thought it would take them until June to flower! I thought they would be so boring!

Again, the vase passes hand to hand around the tables in our windowless stuffy classroom. Noses bent close to the blossoms.

Breathing.

I am stopped.

Uneasiness interrupts serene lesson plan flows.

This image: Faces bowed over the vase.

This image: Breathing.

This image: Me, the teacher, piling reading after reading after work upon work on them for what after what?

Was the breathing not enough? Was the living experience of these growing bulbs not enough for us to contemplate together? And then the moment passed. The danger of it becoming just another 'activity' or "learning object" in our rush to the future was very real.

Part 2—The Time of the Institution

What does it mean to live well in this place, now? The word *now* hangs on the end of the question, positioning the question not only materially in place, but also in time. Other time-bound questions might arise: What is most urgent? What matters most right now? What is the right thing to do for this child, for this planet, for ourselves, for this place?

Educational institutions are almost always rushing ahead of themselves. The industrial, capitalist line of time we unconsciously dwell in much of the time marches us into the future at a furious and inhuman pace, consuming our lives with its promises of progress and forward motion. Stealing the time of the present. We are always rushing but never arriving. A grueling line. A grueling life. I am tired. In *The World We Have*, Thich Naht Hanh (2008) suggests that we have become both victims and slaves to this time.

Competition, measurement, bottom lines, accountability. Preparing for the future overwhelms. Distracts. Imprisons. Enslaves.

We forget *what* we are. We forget *when* we are. We forget *where* we are.

Part 3—Teacher (Preparation) Education

The students lash out at me, angry that I (we) haven't taught them enough. They do not feel ready for the future. They do not feel they know enough. They do not feel prepared. I am not teaching them what they feel I am supposed to. They sense something coming: That time they will be in their own classroom and not know what to do. How can I tell them that they do not know yet what they will need to know then? It's an impossibility. A koan.

Prepare stems from the combination of two Latin words, '*pre*' meaning before or ahead of time, and '*parare*', meaning to make ready. Thus to 'prepare' literally means to make ready in advance of something happening.

Might the proper pedagogical question be *what is it that we are getting ready for in advance?*

Part 4—The Paperwhite's Lesson (Plan)

The deep green stems of those paperwhites are now thick and strong and 18 inches tall. Thousands of tiny white roots wrap themselves around the bottom of the glass vase. The blossoms reach towards the hermetically sealed window in my office that separates them from the elements, away from the fluorescent lights, towards the real sunlight that gives them energy and life. The paperwhites have a sense of belonging—and it is not in this place.

I am a slow learner but I realize now that the paperwhites have become the teacher and are offering up their own pedagogical wisdom.

The scent of the delicate white blossoms is not delicate, soft or subtle. It greets me as the elevator doors open to the hall. In my office it overpowers, overwhelms, almost too much to bear in this small space.

> *My senses are opened.*
> *My senses are opened.*
> *I become aware of my breath.*

In this breath, we are present to the paperwhites, and so present to the breath of the planet. A brief inhaled moment of potential awakening. Can we hold ourselves here?

The awareness of breath is the heart of meditative practices. Aware of the senses, waking up to the world. Conscious of our bodies, of this place, of this time.

Being here. And here again.

The miracle of the bright green of the stems and leaves is the power of the sun to transform a seed into food, to exhale the breath of life. The strong white roots remember participation in these great cycles of flowing water for over four billion years. This miracle calls forth a mutuality. A common planet. A common breath.

The paperwhite is not worried about or planning for the future. It is alive right now. It is living in its own time, which is the only time there is. The only relationship we can have with it is in the time we share—which is the time we are living in now. The paperwhite cannot be in some other time or some other place. It can only be right here. Where it is. Now.

The flower is not afraid. It is not rushing around, nor is it wasting time. It is not filled with wishful thinking. It does not perform to a specific standard at a specific time. It is not in the future.

Might we hold the life of the paperwhite, still, in our hearts?
Might we hold our own lives and the lives of our students, still, in our hearts?
The time and the body of the earth, still, in our hearts?

With this plant we might learn to wait. We cannot hurry it. It is teaching about change and inevitability. It is not anxious that it won't be ready in advance for what is coming. We do not need to prepare it the way we think to prepare children in kindergarten to survive grade one, or prepare children in elementary school for junior high. We wouldn't ever talk this way about this plant, so why about children? Why about pre-service teachers?

We might experience joy in its growing. In our care for it. In our embodied and sensual experience of the miracle of its very life. Of the way it becomes taller each day, the pods containing blossoms ready to burst. And then one day, there they are.

We trust that the flower will know what to do.

The paperwhites might bring the life force of all other plants to mind, unveiling a relationship of profound interdependence that requires awareness and care of much more than our own concerns, that invokes a true and real accountability that can never be measured. That invokes a love for life itself, for the continuance of life on this planet including that of our own species. This concern is also for ourselves and those who are with us, sharing this time, sharing this place.

The paperwhite is a mystery and a miracle. It cannot ultimately be known through any amount or method of study. It has its own subjectivity. And yet our intimate shared breath reminds that we are not and cannot be separate from this plant, as separate egos going through life without 'accountability' to, in Buddhist terms, 'interbeing,' or in ecological terms 'intermingling,' with other life forms and beings through time. It is here and now that the future emerges, whether we are awake or not.

And just as surely as these bulbs sprouted three weeks ago when gifted sunlight and water, this cycle will soon end. The paperwhites will teach the narrative of all life, about the rhythms and cycles, beginnings, endings, renewals and decay in which we also participate.

The paperwhite cannot stop this process.
It just is.
It is being. Being itself.
Being a paperwhite, moment by moment.
Quietly breathing the infinite universe.

Part 5—Coda

Breath of water. Wind of life. Reaching for the sun in these windowless classrooms. Around the circle, hand-to-hand, crinkling brown dry skin invokes language while fingers point to tiny white shoots already emerging, whispering a promise of a plant to (be)come …

12. "The Memories of Childhood Have No Order and No End": Pedagogical Reflections on the Occasion of the Release, on October 9th 2009, of the Re-Mastered Version of the Beatles' Sergeant Pepper's Lonely Hearts Club Band©

DAVID W. JARDINE

"The memories of childhood have no order, and no end."
from Dylan Thomas's (1954a, p. 8) "Reminiscences of Childhood"

We are no longer able to approach this like an object of knowledge, grasping, measuring and controlling. Rather than meeting us in our world, it is much more a world into which we ourselves are drawn. [It] possesses its own worldliness and, thus, the center of its own Being so long as it is not placed into the object-world of producing and marketing. The Being of this thing cannot be accessed by objectively measuring and estimating; rather, the totality of a lived context has entered into and is present in the thing. And we belong to it as well. Our orientation to it is always something like our orientation to an inheritance that this thing belongs to, be it from a stranger's life or from our own.

From Hans-Georg Gadamer (1994, p. 192), *Heidegger's Ways*

Reminiscence I

All this is certainly more than twenty years ago today. Grade 11 English class, Winter 1967, some time after February 2, 1967.

I'm mentioning this because I still remember that date, February 2, 1967, was when I first heard the latest Beatles single, Penny Lane/Strawberry Fields Forever, over an American AM radio station that crackled its way to southern Ontario, to Burlington, one evening whose details are lost to memory. Was it WBZ from Boston? Or that Cleveland station perhaps? And how we stood in Doug's rec room, amazed, wondering whether we were actually hearing what we thought, or whether the radio was blurring in and out of the winter snow-static, like old TV reception.

"Good Vibrations" the previous Fall of '66 was strange enough. Now *this*?

What *is* this?

Part of it was the vertigo tremble of being near-17 in oh-so-fortunate days such as those and sensing all around secret and hidden worlds impending. Glimmer of a sort of *Aufklarung*—with its double German roots of "enlightenment" and also "clearing up."

Some lit cracks "break forth" (Gadamer 1989, p. 458) in those clouds of winter static. Openings. Opportunity. Possibilities, and "thereby possible ways of shaping our lives (Gadamer 1986, p. 59).

There. "Through the recess, the chalk and numbers" (Wilson & Parks 1966). Something secret. ("Yo-da-lay-ee-who," sung right over that song-line).

Some time in late February that year, in our Grade 11 English class, we were given "Reminiscences of Childhood" by Dylan Thomas to read with the horrifying and boring prospect of discussing it in the days to come. (School not yet ever experienced as a place of *schola*: "leisure," and a place of "a holding back, a keeping clear" [OED]).

Unexpected, then, in reading this for school, this happens. I seem to already know intimately of this place of reminiscence. I "recognize [myself] in the mess of th[is] world" (Hillman 1983, p. 49). How did this Welshman know this of me? Was I spotted unawares when I was lingering there? (*Aufklarung* is not simply becoming conscious of threads in the world, but becoming, in all its myriad, *self*-conscious—not just spotting but being spotted, and spotting that, experiencing that I am experienced by others, that I am visible).

Dylan Thomas has experienced me and now seems to write down heretofore secret trails of what I am. Images of flying, in memory:

> over the trees of the everlasting park, where a brass band shakes the leaves and sends them showing down on to the nurses and the children, the cripples and the idlers, the gardeners and the shouting boys. (Thomas 1954a, p. 8).

And these specific streets, "Inkerman Street, Sebastopol Street" (p. 8), names whose very specificity made them both tangible and imaginable.

Nearby recesses.

My self being formed right there, in front of me. Like bankers, sitting, waiting, midst pictures of possible trims, straight-razor foamed clean-cut around the ears.

And there, a "fireman turning off the hose and standing in the wet" (Thomas 1954b, p. 15) (safely rushed in from the pouring rain), from "A Child's Christmas in Wales" (which I quickly read next back then and not at all for school).

Very strange.

Nurses, imagined in memory with poppies in a tray in the park with old men and brass bands, my own old "dame school . . .so firm and kind smelling of galoshes, with the sweet and fumbled music of the piano lessons drifting down from upstairs to the lonely schoolroom" (1954a, p. 7). Piano keys will yellow stained ivory like the reeky fingers of smoky old men or the teeth of piano teachers sat upright and prim. Lakeshore Public School, built in 1923, red brick, facing Lake Ontario, "the singing sea" (1954a, p. 8) where by Grade Six we'd smoke stolen cigarettes and meet at the water's lap edge, a "world within the world [into which we are drawn]" (1954a, p. 4), "secret" (1954a, p. 4), housing "its own, even sacred, seriousness" (Gadamer 1989, p. 102).

"A space specially marked out and reserved" (p. 107).

"This is for *us*, not for the 'others'. What the 'others' do 'outside' is of no concern to us at the moment. We are different, we do things differently" (Huizinga 1955, p. 12):

> And a packet of cigarettes: you put one in your mouth and you stood at the corner of the street and you waited for hours, in vain, for an old lady to scold you for smoking ["a world wholly closed within itself, it is as if open toward the spectator in whom it achieves its whole significance" (Gadamer 1989, p. 109)]. (1954a, p. 17).

Good heavens, there it is: "the wild boys innocent as strawberries" (1954a, p. 6).

What am I going to do with this secret? ("Every game presents the [one] who plays it with a task" [Gadamer 1989, p. 107]).

I ended up, of course, humming along in our Grade 11 class as I read about hunchbacks and how "the boys made the tigers jump out of their eyes," (1954a, p. 6) and "the smell of fish and chips on Saturday evenings" (1954a, p. 7) (tethered later to even more dirty-secret Beatle pies).

And Wally, our English teacher, asking me about the humming song, and me telling him about Penny Lane and Strawberry Fields and how this Reminiscence we were reading sounds so familiar, and him getting one of those purple smelly Gestetner™ machine mimeograph sheets and telling me to take it home and type out the lyrics.

And I did. And he copied it out. And our whole class looked there and then and long and hard into the deep face of the world that some of us were actually already secretly living in.

Ear whisper. Remembered thus, our teacher to me: "You need to read more Dylan Thomas." This is what he said in truth but it may not be what he actually said. I can't quite remember but I remember it as clear as day, that invitation out of yet back into childhood sleepiness, out of and yet back into to play.

And then.

Four months later, June 1 1967, a month to go before school is finally out, K. P. and I go into Hamilton's West End, Kensington Mall (a name worthy of Wales or Liverpool) to a record store, and buy *Sergeant Pepper's Lonely Hearts Club Band*©. Not right away, but slowly, soon, there, some recognition in that crowded cover photo, right above and between Marlon Brando and Tom Mix, themselves above the Madame Tussard wax figures of The Beatles already reminisced from 1964.

There. A round and unmistakable puffy-tousled face in black and white.

There!

Dylan Thomas.

Reminiscence II

> We do not understand what recognition is in its profoundest nature if we only regard it as knowing something again that we know already—i.e., what is familiar is recognized again. The joy of recognition is rather the joy of knowing *more* than is already familiar. In recognition what we know emerges, as if illuminated, from all the contingent and variable circumstances that condition it. (Gadamer 1989, p. 114).

The joy of recognition. The joy of knowing *more.* In blurry retrospect, this is the first time I remember having the experience of actually *knowing something*, of actually having read something "right" up out of my own according life.

The pedagogy of this is important and almost intractable.

"Over and above our wanting and doing," things sometimes "happen to us" (Gadamer 1989, p. xxviii). This, of course, is an insight long in coming and points to why I still remember this event, still call it to mind and make something of it, again and again, in such callings. "Every repetition is as original as the work itself" (Gadamer 1989, p. 122). That there are adventures in the world, secret places that will yield those secrets if I care to venture and do what the place requires of me when I arrive. Our teacher seemingly "ignor[ing] the orthodox who labor so patiently trying to eliminate the apocryphal variants from the one true text. There is no one true version of which all the others are but copies or distortions. Every version belongs" (Thompson 1981, pp. 11–12).

A conspiracy, a common breath.

It is as if I was, all those years ago, presented with a task that was wedded to a prospect, a promise, that settling here will bring unexpected yield, that staying here has a future (linked somehow to reminiscence, to a past's future that is remarkably "standing" still "in a horizon of ... still undecided future possibilities" (Gadamer 1989, p. 112). Still. This is how stillness anticipates. *Stillare*—"to drip, drop" (O.E.D.).

Every version belongs. "Interweaving and crisscrossing" (Wittgenstein 1968, p. 36). Undergoing an "increase in being" (Gadamer 1989, p. 40) in such interweaving. Dylan Thomas became something more than he was and, in such becoming, became who he was.

Slowly becoming what it is. Slowly become who I have become. Criss-crossing.

It would be a mistake for me to portray that June 1967 event simply as some thunderbolt flash of some sort of full illumination at the time in the mind and life of a 16 year old. I've since read of Gadamer's teacher, Martin Heidegger, and his oft-used image of lightning, where a territory gets suddenly illuminated and gathers into a place, a *topos*, a "topic" (see, e.g., Caputo 1982, p. 195). This idea of an opening or clearing or field (see Friesen & Jardine 2009) that suddenly opens up and illuminates, has long since haunted me. See, too, Gadamer (1989, p. 21), on *topica* and the young needing images, like nurses selling poppies, for the formation of their memory. All this is of that reminiscence.

Slow emergence. Flash and then slow glowing arrives, but only if I take on as my own the work of remembering. Hermeneutics is a *practice*. An earthly sort of *Aufklarung*. "You should read more Dylan Thomas" as a call to prayer, to practice, with a promise that something will yield through patience and "a continuity of attention and devotion" (Berry 1986, p 32). Hermeneutics as a ecological practice. All the work I've come to do starts here, at this absent origin.

After all, "understanding begins when something addresses us" (Gadamer 1989, p. 299), but it only *begins* there. Reading "A Child's Christmas in Wales" was compelled as a way to continue to remain in the same place and cultivate it a bit more, as is writing this paper, a gathering that is also a whiling (Jardine 2012e), like the festive (Gadamer 1989, pp. 122–3) 09–09–09 release of the re-mastered *Sergeant Pepper's Lonely Hearts Club Band*©, its date echoing a white album yet to come back in early-1967.

And, just to foolishly rub this wound, we are dealing with an event which *itself* was *already* about memory and reminiscence, in and of Wales boyhoods, in and of Liverpool streets and orphanages, in sound and shiny brass band accordances in Edwardian suits, young men singing, in 1967, about being 64 (and how I'm now actually a year and three days away—including the terrible knowledge that this dating is always immediately incorrect) and me, at that time, 16 going on 17, beginning to feel full-fledge the reminiscent pull of my own life, my own imagined childhood, now clearly ago, and slowly, slowly, drip-drop, becoming the sort of perhaps-never-was that I can henceforth live with remembering and recognizing thus.

"To reconcile [my]self with [my]self, to recognize oneself in other being" (Gadamer 1989, p. 13).

It has taken years for that event to become what it was.

Read that again.

It has taken years for that event to become what it was. What it *is*—"What *really* happened back then in that English class or that ride to Kensington Mall?"—becomes a more and more trivial question, because the question that drives here is "What has become of it?" Even asking "What *really* happened?" is itself a formative act. This is how memory works, because memory doesn't just store information: it does *work*, and its workings are how it shapes the one remembering. This is why memory is always *someone's*:

> Whoever uses his memory as a mere faculty—and any "technique" of memory is such a use—does not yet possess it as something that is absolutely [one's] own. Memory must be formed; for memory is not memory of anything and everything. One has a memory for some things, and not for others; one wants to preserve one thing in memory and banish another. "Keeping in mind" is ambiguous. (Gadamer 1989, pp. 15–16).

Indeed it is. This ambiguity hides an often-secreted mechanism of becoming *myself* right at the very moment when contingent and variable circumstances both fall away and gather up into illuminating recognition all at once.

Look there. Wild boys. Innocent. As *strawberries* (that passage only noticed this time through decades later, reminiscing about reminiscences).

Ambiguous, this keeping in mind.

So, that "past" event's eventfulness is recurrently experienced as "a task that is never entirely finished" (Gadamer 1989, p. 301). After all, here we are, writer and reader, so many years on, still struggling with how and whether to learn to live with the entreaties of this event. "By forming the thing [I] form [my]self" (Gadamer 1989, p. 13), and by informing, now, in this writing, I set out for readers a gentle pedagogical demand on thinking, on memory formation and its ways.

Reminiscence III

"Something awakens our interest–that is really what comes first!" (Gadamer 2001, p. 50). Something *awakens.* Something *clears. Aufklarung.*

> The lane was always the place to tell your secrets; if you did not have any, you invented them. Occasionally now I dream that I am turning out of school into the lane of confidences when I say to the boys of my class, "At last, I have a real secret."
>
> "What is it—what is it?"
>
> "I can fly."
>
> And when they do not believe me, I flap my arms and slowly leave the ground only a few inches at first, then, gaining air until I fly waving my cap level with the upper windows of the school, peering in until the mistress at the piano screams and the metronome falls to the ground and stops, and there is no more time. (Thomas 1954a, pp. 7–8).

You see, I could have sworn, starting this paper, that Dylan Thomas' nurses in the park had poppies.

They do not.

And that there was an orphanage in Wales, wasn't there, like Strawberry Field in wild boy days in Liverpool?

No.

Thus I, too, have a secret.

> A new idea is never only a wind-fall, an apple to be eaten. It takes hold of us as much as we take hold of it. The hunch that breaks in pulls one into an identification with it. We feel gifted, inspired, upset, because the message is also a messenger that makes demands, calling us to … fly out. (Hillman 2005, p. 99).

And when they do not believe me, I write my way down lanes of confidences.

A "Have No End"

So, here's to miraculous happenstance, the remembered ventures of a teacher, and the great lack of assurance that underwrites pedagogy at its best. I know that this age-old experience of being drawn into a world and finding myself already there is one that has been and remains a steadfast refuge and path of my own work, and a key to what I hope can happen in schools.

Here's another small secret that isn't much of a secret. It's not just a hope. I've seen something of it happen in many dame schools, firm and kind and full of fumbled piano music.

At least, that is what I remember through this recess of chalk and numbers.

And when they do not believe me, I flap my arms and write out of the joy of recognition.

It has allowed me to still feel a bit of a wild boy, still somehow innocent as strawberries.

Bragg Creek, Alberta, December 21–31, 2011

13. *Losing Wonder: Thoughts on Nature, Mortality, Education*

JACKIE SEIDEL

> Were our hopes to rely on perfect beginnings and ends, this would surely be cause for despair. But if hope, instead, is our messy, multiform continuance, then what we need is rather to mourn and laugh and dance until our flesh remembers how the world goes on.
>
> Kathleen Sands (in Keller 1996, p. 134)

> When the ground is what the blood knows, we simply greet it and rest.
>
> Cold Specks (from "Lay Me Down", 2012)

Beginnings

Some years ago, while living at the Atlantic coast and teaching in a small university there, I began to struggle with my teaching and research work. I spent a lot of time walking by the surf-crashing, pebble-tumbling ocean. Vast, dark, incomprehensible, different every day. Sometimes calm and playful. Sometimes terrifying and powerful. These water molecules from planetary beginnings had travelled on currents and tides for billions of years, spraying into the air, evaporating and falling again, creating thick, dripping fogs. The shoreline changed each day, imperceptively and sometimes dramatically, sand bars moved, cliffs collapsed, flotsam washed up on the tides. *What is a human being* here on this beach, where life itself is forever being swallowed by long time and great space and some kind of environmental grace? Yet while perhaps this great ocean *seems* somehow wild and timeless, and even the huge international oil tankers anchored off shore appear miniscule and insignificant as they swing back and forth with the massive tides, this moon-pulling, sun-sparkling, mysterious and powerful water that has the power to feed populations, crush rocks into sand, create weather, shape continents, and sink ships is revealing another story. The collapse of the wild salmon stocks or the approaching extinction of the North Atlantic Right Whales in the bay or the unusually high cancer levels amongst the people are a daily reminder to those living nearby of the ecological implications of this industrial time. At the same time, the crushing economic struggles and the clamouring political and economic voice of 'progress,' growth, and the need for more robust economic competition, generated feelings of frustration, desperation, time and space being squeezed shut, breathless rushing amongst the local people.

The nature condition, the human condition, the teaching condition: these were not separate in this place. My teaching and research became trapped in a dualism from which there seemed no escape. In my pedagogical practice and writing I was consciously attempting to pay attention to what was earthly and mortal, yet the conditions and purposes of the work seemed to ask me to ignore limits, relationships, and the obvious collapse of the local ecosystems in favour of promoting a narrow view of literacy and learning. In fact, the social and political undercurrent implied that 'fixing' the perceived literacy problem in schools would 'fix' all other problems. Intellectually recognizing the sources of this dualism offered no comfort in the actual day-to-day practice of pedagogy and research.

If human faith in progress, especially technological progress, creates "an impatience and even disdain for life, a contempt and defiance of our bodily, that is mortal, earthly existence" (David Noble cited in Fisher, 2002, p. 156), what does this mean for living and learning in schools where we are distracted by preparing children for the so-called 'real' world of workers and economic growth? This view which envisions children as resources and subservient to the future human world implies a necessary separation or forgetfulness—at least symbolically—from the material and natural world from which all bodies and spirits always have been born, sustained, related, and to which all return. The distracting, distancing vision of preparing for the future, of rushing ever forward, may create a sense that our own mortality will never come and can always be postponed, we can behave as if we, and those we teach, are immortal. At least for the time being. But when these tensions and forgettings are ruptured by the present, by terror, politics, illness, or simply the struggles of day-to-day human relating, and we are brought to *face* our shared earthly mortality it might become literally impossible to continue with confidence certain educational practices and ways of being.

Rosemary Radford Ruether (2002), in reflecting on these dualistic thought patterns in the West, encourages a reshaping of "our basic sense of self in relation to the life cycle" (p. 22). In meditating on the profound interconnectedness of life and time, we might remember that our bodies, each one, are also recycled, like our garbage, like our breath, like the calcium and iron and salt in our bones and blood. In the ocean. This wondrous flowing moving intimate kinship with life. Rosemary Reuther continues:

> The western flight from mortality is a flight from the disintegration side of the life cycle, from accepting ourselves as part of that process. By pretending that we can immortalize ourselves, souls and bodies, we are immortalizing our garbage and polluting the earth. (...) Humans also are finite organisms, centers of experience in a life cycle that must disintegrate back into the nexus of life and arise again in new forms. (p. 22).

What is pedagogy that can bear the wonder and mortality of the world?

Meditation One: River Walk

The North Saskatchewan River flows through the Canadian prairie city of Edmonton. Walking paths lead for many kilometers along the river's banks. The water is cold and wide and deep, curving broad banks along deep forested ravines, under the downtown building-jutted skyline. Born hundreds of miles west, high in the Rocky Mountains, the river begins as a slow ice-melt trickle. Under the glaciers, thousands of years of compressed snowfall gathers momentum, moving and grinding sand and pebbles. At the source, the icy water is bubbling shallow and bursting fast. Further on it deepens, widens, as many other waters join it, carving out a wide, curving valley through the prairie, supporting a huge ecosystem for more than one thousand kilometers. Humans interfere, now and then, here and there, taking water out, putting it back, damming and changing the flow. This river was the 18th century colonizing route into the 'west' as map-maker and explorer David Thompson built a system of fur-trading forts and mapped out the land for settling Europeans. A carrier of story, this river also holds the "deep time memory" (Kostash, 2003, p. 52) of Aboriginal peoples—the Peigan tell of the glaciers receding 12,500 years ago as the Creator walked north. It has taken time for this river to draw its path here. But the geological history of this place exists in deeper and longer time, following the flow of least resistance in its yearning to join the other waters. This huge river is but a tributary, many waters flowing together and merging with the South Saskatchewan River to become the Saskatchewan, flowing eventually into Lake Winnipeg and draining towards Hudson's Bay. Where does this cycle begin and end, this endless flow of water molecules around the planet, through life?

One warm, but not-yet-too-hot sunshiny day, leaves still that pale green rustling, I walked near the river. The smell of clover. The sweet silvery wolfwillows on the banks. The poplar trees leaning over. A flash of pale blue in the undergrowth beckoned. Crouched low in the woods, springing up lovely from the forest floor, the clematis was blooming. I was glad to see it. Its petals tissue-paper fine, purple-blue and near translucent, yet its stem is strong and stable in the breeze. In a few weeks it would be gone, transforming into miraculous fuzzy seed heads loosening their seeds into the rotting undergrowth that is already preparing for the wild prickly pink roses, fireweed, dogwood, new tiny trees to come.

As I crossed the old train bridge, two young boys leaned over the rails, looking down at some fly fishermen standing in the flowing water. *Are you going to eat those fish?* one yelled. They looked at each other and start to laugh hysterically. The other yelled down, *Gross! When I flush my toilet it goes into this water! I sure wouldn't eat those fish!* They laughed some more. Ignoring the boys, the fishermen threw their lines into the water.

When I was eleven years old, about the age of those boys, my family moved from a large city to a lake in a provincial park west of the small town of Rocky Mountain House. Encountering the blooming clematis flooded my body and senses with memories. My child mind awoke and remembered those first heady days of my new rural life. The smells! The frogs! The insects! The forest! The sandy-bottomed lake full of fat leeches and small minnows and hair snakes! It was all marvelous, endless places to explore and play. My new grade six teacher assigned us a plant study project. We were to find a flower or plant, identify it from books, press and dry it and share it with the class. It was a sun-burnt sappy pine day, warm pungence in my nose, the leafy light filtering layers of spring speckling as I walked and searched for a plant for my project. And then, there it was, the wild blue clematis. It was perfect. Beautiful. I learned its name from my mother and took it to school the next day.

The teacher showed us how to dry our flowers in salt, and several weeks later we removed them carefully—they were even more fragile now—and we pressed them onto cream, coloured cards under clear plastic sticky film. The clematiss tough woody centre made it difficult to flatten like we were supposed to. One of the leaves got bent over under the sticky plastic film and there was no way to pull it off. I decided that I liked how it looked. It seemed natural somehow. I loved how the folded petal meant that I could see both the front and the back of the petals on my card. When the teacher asked for our cards, I handed mine in, sure that he would be happy with it. Sure that he would be happy *with me.*

Our flowers were returned to us with a rubric. It is the first time I ever remember receiving a grade at school, although I must have gotten them in previous years. My brain was dizzy. How could I only have gotten 4 out of 10? Didn't my teacher like my flower? Didn't he like me? Marks had been deducted for the petal having a fold in it and I can't remember what else. Only the feeling of having the wonder of the experience crushed. The smell of the woods, of choosing the perfect flower, taking the assignment seriously, the crunch of branches and dried grass underfoot, the singing of birds. The assignment had made me love this flower, provoked me to explore my new home, opened the wondrous forest floor to me in new ways. My whole body, city-child turned country-child, had opened to the full, sensuous, complex beauty of the forest. I had experienced the woods with awe, perhaps for the first time in my life with that *wonder beyond words.* My relationship with the clematis extended through the whole forest, through all my senses. It became part of me. None of these things could be measured, and so a full and wondrous experience was instantaneously flattened, reduced to a folded over leaf failure. The teacher's response had exposed me to a pedagogical and social violence that I don't remember experiencing before that. It changed my feelings about school. The sensuous experience of encountering this flower—the clematis and my relationship with it—could not co-exist with the weight of the teacher's narrow rubric or our stuffy basement classroom. With a few lines on a paper, the teacher completed our (op)pressing of the flower, objectifying it, severing it from the fullness of its surroundings. From the tall pale-barked poplar trees, the sun dappled earth, the sweet smells, the buzzing insects above, around and below.

After so many years, in this new city, along the same river that flows through Rocky Mountain House, I was glad to have encountered this clematis again. My body and senses remembered the forest floor, the cool air, the smells. I was even glad to remember the teacher's response to my work, to have this chance to contemplate the meaning of this pedagogical work, and to be reminded to be forgiving of all of us, of the things we do as teachers, all those well-meaning and hopeful things like rubrics that we think make our jobs more organized and simple. I was glad to be reminded of how unaware I might be of children's relationships and connectedness to the world, how easily I might crush a child's experience and wonder without ever realizing, how easily I might violate a child's relationship and trust in me as their teacher, that I might engage them deeply in a conversation with the world, and then rip it from then again in the name of mindless institutional processes.

Questions come with these memories, haunting this walking path now. Philo Hove (1996) reflects that wonder itself "lies at the heart of what it is to be human: it places us directly and transparently in the face of the world in which we live with others" (p. 437). How might we as teachers share this wonder in our relationship with children? Learn to recognize it in the other and let it be, let it form and grow? What if it matters, in this time of great extinctions and global human suffering, what if it matters, to the future of life, that we learn to do this?

Meditation Two: Re-Reading Crow Lake

Mary Lawson's evocative first person novel, *Crow Lake* (2003), set in a small northern Ontario farming community and Toronto, circulates powerfully around themes of family, home, community, and identity, and particularly how these relate to the death of the narrator Kate's parents when she was a child. Written as a memoir, the story balances between Kate's early life, after her parents' sudden death in a car accident, when her older brothers gained custody of the younger children, and her life as an adult when she has become an ecologist studying the effects of pollution on freshwater pond invertebrates. Particularly significant to the story is Kate's relationship with her brother Matt, who often took her to the ponds near their home to learn the intimate exchange of life there. In sharing this place with Kate, he gifts her with his passion, experience and knowledge. Barely an adult himself at the time of their parents' death, Matt becomes not only Kate's caregiver, but also her teacher.

My interest in this story centres on a narrative turning point when the adult Kate experiences what she calls "a bit of a crisis at work" while lecturing at the university. She describes it:

> Anyway, this 'crisis,' if that isn't too dramatic a name for it, came in the middle of a lecture. It started as a minor hiccup. I'd been explaining the hydrophobic nature of the hair piles of specific arthropods to a lecture hall filled with third-years, and I suddenly had such a vivid flashback that I completely lost my train of thought. What I remembered was Matt and me, in our usual pose, flat on our bellies beside the pond, our heads hanging out over the water. We'd been watching damselflies performing their delicate iridescent dances over the water when our attention had been caught by a very small beetle crawling down the stem of a bulrush. (p. 197).

Derailed from the purpose of her lecture, and awash with overwhelming memories of the embodied life of the pond so different from the picture on her overhead, Kate's vision is again interrupted by a student in the front row of the lecture hall yawning widely. Kate continues:

> I stood speechless, staring out over my audience. Inside my head, my inner ear played back to me the sound of my voice. The drone of it. The flat, monotonal delivery. And overlaid on top of the drone, like a film joined up with the wrong soundtrack, I kept seeing my own introduction to this subject: Matt and I, side

> by side, with the sun beating down on our backs. The beetle sauntering along under the water, safe in his tiny submarine. Matt's amazement and delight.
>
> Matt thought it was miraculous—no, there is more to it than that. Matt *saw that it was miraculous.* Without him I would not have seen that. I would never have realized that the lives which played themselves out in front of us every day were wonderful, in the original sense of the word. I would have *observed*, but I would not have *wondered.*
>
> And now I was putting an entire class to sleep. (p. 199).

In this moment, Kate, the professor, researcher, expert in her field, suffers a kind of psycho-spiritual breakdown. She gazes at her class, silenced, completely unable to continue the lecture. Apologizing to her class for boring them, she exits the classroom.

When reflecting later on this 'crisis,' Kate says that although she is able to return to the classroom for the next lecture, she is now utterly exhausted by teaching. She diagnoses herself as being "sick at heart" (p. 235), experiencing a profound sense of disconnection between her embodied knowledge and her physical reality as a university teacher. She is dislocated, ill at ease, and suddenly not at home in her classroom or at the university. When, throughout the novel, Kate relates the memories of Matt sharing the ponds with her, she describes the deep experience of existential wonder, her intimate relationship to her brother, and her physical, intellectual and spiritual rootedness in the place where she grew up. But then, the yawning gap of her student's mouth knocks her off her confident bearings as a professor. She realizes that there is no connection between herself, the students, and the 'knowledge' that she was supposed to be giving them. As Matt had given her. As a teacher she is destroying something that had been a source of love, wonder and life-energy for herself. Kate is, perhaps literally, *killing* her subjects. She has turned Matt's miraculous creatures into sterile, knowable images on plastic overhead sheets, rather than fragile and miraculous living creatures of ecosystems. Life, in her classroom teaching, has become objectified and knowable, but her memories are telling her otherwise, cutting open the sterile space of the university classroom. Her memories are the ecology of her life. The water beetle they study is intimately connected to her many hours of wonder at the ponds with Matt, the death of her parents, her home in the north. None of it fits in this room at the university and she loses her voice, her ability to speak in any meaningful way.

Where did the wonder go, in the space-time between the pond and the classroom? In the distance between her relationship with Matt and the ponds, and her relationship with her students and their learning? Philo Hove (1996) brings attention to the experience of wonder in human living: "Attentiveness to wonder, and to the many dimensions of experience it reveals in our lives, can cultivate a sensitivity to the emergence of wonder in others, and therefore, has significant implications for the way in which we can be pedagogically oriented towards students" (p. 437).

As a professor, in a university science building where walls create physical distance, where the city creates physical distance, Kate is unable to connect back to the wonder she once felt about these pond invertebrates, or to her relationship with Matt. Her relationships are dislocated by the situation and she is utterly lost. Rather than being able to *"cultivate a sensitivity to the emergence of wonder in others,"* her voice is silenced by a student's yawn. Any possible pedagogical orientation to her subjects or students seems trivialized by the institution and practices of the university. An insect, which belongs to the world, when encapsulated on an overhead projector and in words, paralyzes Kate. What learner or teacher does not know this jarring and dislocating experience? In these tired, yawning classrooms, is it even possible to experience relationships? Or does the institutional structure preclude, erase and make impossible the kind of pedagogically oriented relationships that Philo Hove describes? "In a world without wonder there is nothing to enter into relations with; because the world is mute, colourless and inanimate, we lack the means for really living in it" (Hove 1996, p. 441). Although she is 'close' to her pond invertebrates in her lab, in tanks at the university, in her teaching, Kate feels alone and far from anything that *matters*. In spite of all her hard work and becoming an 'important' ecologist and academic, she has essentially lost her 'home' (significantly, the root of ecology is *oikos*, home).

Kate becomes convinced that her teaching crisis is related through time and space to all that unfolded around her parents' sudden and violent death, as a life event for the siblings and the community, into the flow and becoming that has been her life and work. Kate achieves insight into Matt's relationship with the pond when she says, "I'm sure he drew comfort from the continuity of life there. The fact that the loss of one life did not destroy the community. The fact that the ending of life was part of the pattern" (Lawson 2003, pp. 102–103). But in her crisis and aloneness, and with her own research focusing on the effects of pollution on pond life, Kate lacks faith in the continuity of life. She is worried sick, afraid of going "home" to the ponds. Imagining extinction, she says: "I imagined myself, going back to them one day in the future, looking into their depths and seeing ... nothing" (p. 190). Kate's academic research—her educated 'knowing'—have come at great personal cost. She has lost her home/*oikos*, her heart, her wonder, her sense of relatedness, unsurprising perhaps to anyone who has ever been in a university where the 'methods' are so often about objectifying, and separating the heart from the mind. Without wisdom, 'knowing' has no purpose other than creating more knowledge (but for what? serving the knowledge economy? constructing learners as workers?). Kate's seemingly perfect professorial life comes crushing in on her in a silencing exhaustion.

In her moment of crisis, Kate unconsciously understands that a pedagogical relationship is not being formed with her students, that she can never share 'knowing' in the traditional way of standing at the front of the room and illustrating life on an overhead slide. The chasm created by the classroom and the institutional structures means that the subject has become only an object for the violence of their *learning* gaze. Kate's silence is, in this way, perhaps a healing gesture, the only proper response, the only way to do no more violence.

Kate's inability to speak in her classroom prophetically whispers to us as teachers, perhaps stirring up quiet questions and terrors, and causing us to wonder, but not *wonder*, at what we are doing in this work, at what is the institution asking us to do. Have we lost our home? Our voices? Our selves? The possibility of ethical pedagogical relationships? If we understand that the self only exists in relationship to the world, then the severing of relations is a kind of extinction of self, a profound emptiness that can never be filled by literacy programs and other busy things unrelated to the ground of our life being.

Kate's story, for the meaning of teaching, however, is not hopeless or pointless, but rather *hopeful*. When Kate suddenly does not recognize her place or self in that classroom, this 'crisis' leads to her going away from the university for a while, to reconnect to those places that matter to her spirit, to re-establish the severed relationships and wholeness in her world, and to come to terms with the story that is her life. Her story points to the possibility (and inevitability) of awakening to the disaster that our educational institutions can be, on an individual psychic and spiritual level, but also on the level that they participate in creating the world we live (and die) inside. Kate's journey shows a pedagogical path, a directionality, away from these kinds of teaching and knowing habits, towards more connected, rooted, deep and mysterious ways of being in the world. But can we do this and continue to teach?

I don't know.

Meditation Three: 2000 and None

The Canadian film *2000 and None* (Paragamian, 2000) begins with paleontologist and professor Benjamin Kasparian learning that he suffers from a terminal brain disease. When his doctor explains that he has just a few weeks to live,

Benjamin stares, then stammers: "But what about my research? What about my *finds*?!" The final stage of Benjamin's illness, which he personally dreads most of all, will be the loss of his memory. Pleading with his friends, who are struggling with the 'news' of his imminent mortality, Benjamin makes them promise that when this happens, they will explain to him that he is going to die, to not let him continue unaware of his approaching death.

As a paleontologist and professor, Benjamin life's work has been about exploring and exposing time, excavating the memories of the earth, layers of life and death crushed together, telling the story of life's relationships between past and future. The ending to Benjamin's own life is like an overlay on his work bringing him nearer to the earth, to the fragility of the life (and death) that he digs up every day. The earth, he knows, is full of bones and blood and stories. Despite his initial shocked reaction, clinging to the importance of his own knowing and research, he accepts the inevitability of his own death. His friends, on the other hand, are furious. They want Benjamin to *do* something. As they all share lunch at a restaurant in Montreal, they demand to know how he can eat at a time like this, as if all is 'normal.' They want Benjamin to tell them what plan 'B' is. He looks at them seriously: "There's no plan B ... Life is not just life. It's life and then death. They go together."

There is a critical and pivotal moment in this story. Like Kate, the professor in *Crow Lake*, Benjamin also experiences a self-shattering teaching crisis. In front of a lecture hall full of young students, standing and delivering a detailed and organized lecture at the blackboard covered with chalk diagrams illustrating the layers of the earth, he suddenly gazes into the distance and goes off on a tangent: *"When we think about the layers of traces, of death, of corpses, skeletal remains, fossilized imprints, markings. Life. The comings and goings of one species after another, all there sometimes just metres away but millions of years apart ...."*

Suddenly silent, Benjamin stares at the class, frowning, confused, thinking about what he's just said, and obviously unsure of the rest of his sentence. Like Kate, he has lost his ability to speak about the work that has intimately occupied his spirit. He gazes across the room, at his students, then says, "Excuse me one moment." Benjamin exits the lecture hall and never returns.

What is the meaning of this moment in Benjamin's experience as a teacher? This moment when the truth of his mortal life, the truth of his work, comes crashing in upon his pleasant world of working, going for coffee, enjoying his nice apartment. The expanse of time, of all species that have ever lived and died, faces him, just as his students face him, leaving him paralyzed. What happens, in the sometimes banal contexts of schools, when are faced with *"the layers of traces, of death, of corpses, skeletal remains, fossilized imprints, markings. Life ..."*? Life. All those complexities of relationships over a vast expanse of universe time. In a sense, this statement from Benjamin renders all other words meaningless. What else is there to say? He has nothing else to teach his students. All the years he has spent studying, digging, *knowing*, bring him to this moment of deep understanding that he, with his important 'research' and 'finds,' is not above and separate from these layers of life. Intimately related in this time, Benjamin is not excused from the experience of becoming earth, from what he calls *the comings and goings of species.*

As his health deteriorates, Benjamin becomes afflicted with a myriad of physical and psychic symptoms. Suddenly throwing away his eyeglasses, he declares that he can now 'see clearly.' He hallucinates: silent black and white film memories of his childhood begin to play unbidden in his mind. His parents, dead since he was small, visit him as figments of his imagination and give him advice. He weeps to his mother, "I don't want to die!" She responds: "But you have to." Benjamin becomes obsessed with returning his parents' bones to their ancestral home in Armenia, but his efforts are frustrated at every turn, by bureaucracy, by people who do not understand his intense and sudden urgency. Benjamin's sense of time has shifted. He tells his friends, "*There's no time for adjusting. There's only time for having time.*" He steals into the cemetery at night and excavates his parents' bones illegally, only to lose them somewhere on his desperate trip to Armenia.

Benjamin's father appears to him in another hallucination and says: "*You're more free than anybody.*" His father's wisdom redefines Benjamin's relationship to his own inevitable mortality. And it is this theme that makes the film a profound story about the memory of our own human mortality, as individuals and as a species, of our own place in time. It is a powerful film about life's process (not life's progress). These are lessons that might free us to do good work in life, because *it isn't about us anymore*. We are here, now, and that's all, part of the layers and layers of life. Life slipping away from him, Benjamin cannot hang on to anything. Our self-importance, both collectively as a species, and individually, fades away.

After Benjamin's initial question to the doctor about his 'research,' his 'finds', he never asks these questions again, coming to understand their irrelevance against the background of *layers and layers of life* which have been his real work. Life in this place is no longer about Benjamin, not about his work, not about what is good for him or good for his students, not about his friends, not about the future. His insight comes through his work as a paleontologist. The institution of the university made his life's purpose about his 'finds,' and his 'research', but Benjamin realizes that the ecology of his life, including his grasping for his ancestral bone-filled family history in Armenia, does not *fit* in the lecture hall of the university, flattened by diagrams on the board, or by the bored stares of his students. It is not a separate story from him, but a story he is always and already inside. He leaves the classroom because there is nothing for him there and there is nothing left to say.

Consider that our identity in schools, created on that futuristic model, on measuring and categorizing, on 'making progress,' reveals little to us about our fluid human identities, or who we are, in the complexities and timings of the world. As teachers, like Benjamin and Kate, we are silenced and shattered when *life* brushes up against us in our classrooms, when something moves us to tears or we feel paralyzed in our work. The film's message is the same message we get when standing before the fossils of the Burgess Shale, when an ancient archeological site is excavated, when we learn something new that changes the way we understand everything, when we know millions of our fellow humans are about to starve to death and we are helpless in the face of it, when we read an amazing novel, when nearly a million people are slaughtered in Rwanda, when we are stunned by natural beauty, when the World Trade Center crumbles to the ground, when a child in our class has his throat slit at school, when one nation declares war on another.

When we see a new spring clematis.

These are places of speechlessness, beyond narrow human conceptions of time and progress, and they call into question the 'importance' of the human species, of our own lives, of our neighbours' lives.

A friend and teaching colleague encourages teaching as if all students were terminally ill. *Teach as if they had no tomorrow. What would teaching look like then?* she asks. How would we live together with children? What would the priorities be? How would we address them? Maybe almost nothing we do now in school would seem relevant. In this view, what rises to the surface? Love? Mourning? Memory? Something else, other than this life which consumes us with its goals, objectives, futuristic thinking, preparation for toiling in the global economy, consumer goods, children described as resources. Human Capital.

Robert Thurman (2004) writes that "(t)ruly realizing that your life could come to an end at any moment generates a sense of urgency" and that out of this urgency arrives a "powerful source of energy and a fount of creativity" (p. 199). This is not at all the urgency of industrial acceleration and panic. It is an urgent halt. Benjamin Kasparian's experience, as a teacher and researcher, and human being facing his imminent death puts him in a position of realizing what matters to him. There was only one thing to teach and he said it in a few sentences. *Life.* His own life takes on a sudden urgency. The future is gone and there is no more waiting. He wants only to fulfill his parents' last wish.

This urgency of facing our mortality, instead of an immortally imagined future, *moves* us, not into more hurrying and scurrying, but into doing what matters now. Benjamin's friends complain to him that they have no time to get used to his dying. He tells them "there's no time for adjusting, there's only time for having time." *There's only time for having time.* This urgency of living embodied in time and place, and in the relations we are in right now, rather than in a future that does not yet exist. bell hooks (2000) encourages us, in our work and living, to live in this more urgent place called *now*:

> Understanding that death is always with us can serve as the faithful reminder that the time to do what we feel called to do is always now and not in some distant and unimagined future .... Living in a culture that is always encouraging us to plan for the future, it is no easy task to develop the capacity 'to be here now'. (pp. 203–204).

Being here now. An invitation to a different relationship with time. No more waiting. No time for adjusting. *Only time for having time.* With this attitude we make time more spacious, more open. We give it substance and body—*our* bodies that exist and live and breathe and relate here now.

These stories evoke the ways that the very structure of the educational process allows embodied experiences, memories, relationships and continuities to be easily lost or broken. We are distracted by pressures of accountability, by the way things are supposed to be done, by intense demands for certain kinds of performance—all linked to the 'future' and distanced from our present urgencies and relationships. No wonder we feel exhausted. The intense pressures of our future fantasies and longings erase the layers and layers of life that hold us up, distracting us from our ability to stay here now.

Benjamin Kasparian's question about his 'research' and his 'finds' interests me as a question about teaching identity. In educational institutions it is so easy to think that we *are* our work. We run very fast, working always harder, feeling like life would stop without us. But it doesn't. It grinds on, grinds us up, rips our experiences away from us, and that precisely is the insight that might allow us to intervene with new insights and *a new (other? different?) time sensibility.* Catherine Keller (1996) encourages us to think of our identities as created and dependently bound by all our relations and by the time of our lives: "Time comes tensed, edgy, rimmed by tragedy, edged by all of our deaths: the frame of finite relations which constitutes the moving boundary of any moment, from the complex of relations which I have become to a new (but barely new) complex I have yet to embody" (p. 134). To understand that the person we are always becoming is also being shaped by these institutions might propel us into action, trying to change the institution's identity and character, to make it more humane and generous. To bring it into a deeper and more generous time where we can remember that human (not just teacher) identities are so dependent on not only each moment of our lives, not only our ancestors' bones, but also on the entire time of life's existence.

Could Kate, Benjamin, I, find ourselves at home in our classrooms then?

Meditation Four: On (Human) Nature, Justice, Community

It is a human existential need for our lives and words to have meaning and purpose, not as 'work' or as 'finds,' but as part of the community of life. As they suffer the loss of their existential wonder, as I did with that pressed clematis, these teachers realize that their lives have become separate, dull, lifeless. In the flattening narrowness of their classrooms and university research, in the rush towards immortal futures, the fullness of life in all its complex relations has been lost. In their teaching crises, the essential vulnerability of these educational spaces has been powerfully exposed, and that is cause for hope. This 'hope' does not come from human doings, but rather is the deconstructive movement of life over which no human control can be exerted.

What might it mean, then, for education to hold close to its heart what Catherine Keller (2003) calls our "edgy finitude" (p. 7)? In these three stories about 'normal,' even 'good' teachers, the split between the intimacy of embodied relational experiences in the world and the devastating violence of the classroom on relations and knowledge wounds and silences, paralyzing the possibility of 'teaching.' During my time living at the Atlantic, gazing out at the waters through my apartment's ancient wobbly glass windows, my attention was captured by the nearness of the North Atlantic Right Whales. I never saw one, but they were there in that mysterious grey deep. They will probably be extinct in my lifetime, their own mortality as a species imminent and provoking nearby humans into urgent but helpless concern. While in classrooms with student teachers, even while the windows looked out over the Bay of Fundy, I was conscious of how difficult it was to summon up the presence of the whales there, to make a space for them to exist. *As if already extinct.* Meditating on the condition of the whales is at the same time a meditation on the human condition, on our shared planetary and mammalary existence, dependent always on water and breath. Devastating to my own work and teaching, like Kate and Benjamin, silence fell upon me. My words were frozen. I could not write for an entire year. I did not know what to say, or how to converse with the politicians and school board officials who were demanding more accountability and higher literacy levels at any cost. Even if these costs might be incalculable and devastating to life, individually and historically. The relational and ethical questions provoked by such meditations on our own implicatedness and responsibility to life are very great. If we have classrooms, enthralled by visions of technological paradises, separated from the world by diagrams, overheads, whiteboard imaginaries, where is there space for the intimate cycles of life/death? The classroom shrinks. Filled with so much noise and clutter. Even the talking stops. Maybe the breathing will stop.

Perhaps no amount of effort on the part of my grade six teacher, Kate, Benjamin or me as a teacher, could have brought the ponds, the forest, the layers of life, into those classrooms. What if the conflict in meaning, being and purpose is simply too great? What if such a psycho-spiritual collapse into silence is a sign that this space must be physically abandoned if we are to save ourselves and our students from its violations? Or is the silence a sign, a place to wait and be, to pause amidst the noise and the clutter of our lives? The crisis, the breakdown, a symptom, a turning point opening out into other possibilities? A different direction can be taken and other paths open up before us. Not a cataclysmic end, but rather an opportunity, possible new beginnings, space for difference. Change. Something new might be coming.

Pedagogical spaces are haunted by the unconscious troubles of current ecological and economic crises. What is the meaning of facing the mass suffering caused by modern living, the death of whole systems, ecosystems, cultures, languages? It is a *real* loss, to the *real* complexity of a diverse planet on which our interconnected breath depends. I sense the exhaustion and anxiety everywhere in institutional spaces—the frantic busyness of never arriving anywhere that matters—in my own work at the university, my most recent years as an elementary teacher, and when I visit schools with student teachers.

What of the mortality of seven billion humans, this contemporary mass extinction of species of planet and animals, this finite planet? As a species, we may have reached this point of considering not only the singularity of mortality, a child's, a colleague's, a friend's or our own, but mortality on a much bigger scale. The difficult implications of facing this in pedagogical work are the healing challenge in the present.

Creating Community

In asking which way(s) to turn, philosophically and in practice, consider that community, for which we often say we strive in education, is formed in the recognition of relationships; that is, not in a bunch of unique individuals competing for selfhood, but in the flow of a tangled complexity of interdependence through time. Consider the possibility of living through the silence, the crisis, the speechlessness, in recognizing what it is a symptom of. Consider the possibility of not walking out the door, but of summoning up the courage *to stay* in the classroom and to recognize that all the relationships of life exist there already. They come through the door with each of us. Might we invite their presence, summon up meditative powers to hold them close, against the erasing and demanding powers of the institution? Might we set our hopes on an 'outcome' that is more time-bound, nurturing, meditative, conscious, complex and spirit-filled?

Encouraging us to avoid the lure of both future fantasizing and death denial, Catherine Keller (1996) suggests that it is in the intimate urgency of the present that community is formed, "in the folds of this finitude" (p. 296). She insists that community is reciprocity, is justice, social as well as ecological:

> In us talking mammals the matrix craves conscious reciprocity. This entails rightness of relation: justice toward each other and toward the ecosphere function as *necessary* conditions of this communing complexity. (Keller, 1996, p. 302).

Before both Benjamin's and Kate's silencing, only the teacher is talking, as in so many educational spaces. Both "communing complexity" and "community formed in the folds of finitude" are essentially absent. Because life is ecological and arises out of relationships, Kate and Benjamin's work no longer makes sense to them and the feedback they get is yawns and boredom. It is their own relationship with mortality (for Benjamin, his impending death; for Kate, the memories of her parents' death and the delicate sensitive ecology of the ponds that created her life's work) that spurs in each a deep sense of life's urgency that interferes with the timing and meaning of their teaching.

To learn to be in our classrooms with such a sense of urgency, towards communities of social-ecological justice, so different from the urgency of rushing to the future for the economy, is a challenge for teaching. If we believed this were a measurable and achievable goal we would surely give up; rather, perhaps it is about cultivating awareness, reminding myself to take responsibility for the kind of place I create for children, or teacher education students, practicing living a grounded life as a mortal human being under the idea that what I take up with students in a classroom/curriculum might become compassion, love, life, *how we live now*.

Our knowledge of our finitude, the earth's finitude, is what American poet David Whyte (2001) calls "the ultimate context of our work" (p. 61). Preparing children for the future (and their identity as 'workers'), as a primary purpose of schools, is neither relational nor urgent. The irony is the way it makes us *feel* like rushing! Time stretches endlessly before us and sacrifices our present relationships to future time. In our distraction, we might forget that the present is the only truly intimate place, materializing out of relationships, forming possibilities for the future that we do not yet know. Whyte writes: "Death is much closer to each of us than we will admit; we must not postpone that living as if we will last forever .... All around these conversations, the world is still proceeding according to mercies other than our own. This is the ultimate context of our work. The cliff edge of mortality is very near" (p. 61). Jacques Derrida (2001) insisted that it is our 'knowledge' of finitude that makes possible friendship, relationships, these *communing complexities.* In the classrooms in these stories, in the institutional orientations to knowing, learning, research and 'finds,' the relationships were already profoundly distanced, if not completely severed. Intimacy and love in any form were absent, the classrooms existed in no place and no time. Without community and complexity, the knowledge of finitude was unbearable there and its intrusion precipitated paralysis.

An education that could bear mortality is an education more interested in the natural world, in relationships, in slower and deeper time. It is an education that does *more* in school: more interesting, more challenging, more human, more lively conversation and work. Yet without becoming more *busy*! It might be full of emergent wonder, sublime hopes, and impossible dreams. It is not the stifling, limited, rule-bound talk of accountability to the future, to corporations, to governments governed themselves by corporations, and it does not involve spending vast amounts of money buying more programs for fixing children to fit economic neediness. Open to mystery, an education that 'remembers' mortality is an education holding the relations of life always present, understanding that there is so much about life that can never be 'known.' It might be an education that does not cut itself off from what is outside the school, separating what is called 'knowledge' from bodies, time, experience, continuity, ancestors, but rather being firm on that ground(edness). It is an education that understands that each moment is important, now, bound in a complexity of relationships, intertwined with time, culture, earth, learning. It is not difficult in the current educational context to imagine the opposite, where schools would be (about) the death of time and life. Where love and growth are sacrificed to the future, leading people to be cruel and competitive in their fearfulness and mindlessness. Consciously working against this is not easy or simple. Embracing complexity and finitude, opening ourselves to the relationships that create us all, deeper meaning and time might take root and grow a new kind of place. In the stories invoked earlier, it was the complexity that shattered those teachers in their classrooms. Their work was no longer simple or straightforward, no simple lesson-by-lesson timeline, but rather was interrupted by meaning and time, by nature and mortality.

This Is the Turning Point

> I do not relate to you *in* time, or greet you *in* some pre-established space; rather, our relationship will constitute its own spacetime. Such attention to space brings into the foreground once again and at once the local, the carnal, the cosmic. And therefore at once the temporal, that finitude which allows its members no pretense of their own infinity. (Keller, 1996, p. 170, emphasis in original).

Near the end of the film, near the end of his life, Benjamin Kasparian, once a respected professor whose identity depended on his research and finds, sneaks into the university lab after hours. He steals a brain, bones, some blood. At a place outside the city Benjamin digs a pit in the ground. He lays out the image of a running figure, a brain on legs. After covering this strange grave with earth, he lays down upon it and falls asleep. When he awakens, Benjamin's memory is gone. He has gifted himself, his life, his work, back to the layers of the earth already taking him into its own meaning.

Benjamin's final act is a gift, not just to the earth, but to us as we watch him. It is an act of humility and groundedness, a subversion of the human arrogance that puts us above the earth, reminding us to be careful how we imagine ourselves, our powers, our futures. It reminds us that we belong to the ground.

Rather than allowing these experiences to silence and paralyze us, perhaps we might learn to recognize and respond to these critical symptoms when they arise. To make space for terror, wonder, awe, confusion, those deep human emotions that relate us to the world, to one another, and to time itself. Canadian poet Don Domanski (2002) relates this to his own writing and *being* practices, reminding us that these depths of human life are experienced as a whole, that abjecting one of them (like mortality) might mean also that the rest also disappear from our experience: "My need to write about death is my need to be human. It's not possible to write about life, about the sheer wonder of that, without the backdrop of death" (p. 253).

The moment that Kate experienced in her classroom, or that I experienced as a child receiving a grade on my pressed flower, creates a rift between deeply corporeal experiences embedded in the flow of life, and the present time experience of a classroom separated from any meaning that matters. The characters in the film and novel, and I as a child in the woods, experienced subjective awe and wonder, in place-time-spaces shared with our *subject*, whether a flower, pond creatures, or bones millions of years old. *Layers and layers of life.* Intermingling. Merging. Listening to, observing, and creating such experiences as teachers and learners might be the transformational opening experience we hope for. If, as poet

Phyllis Webb (2002) suggests, "[a]we and wonder bring us close to the mystical, the hope of climbing out of the boxes we live in from day to day" (p. 228), then seeing ourselves *in* nature and *as* nature, as time and in time, as mortal beings, we might learn to resist the "Western preference for future" (Keller, 1996, p. 135).

How is education to bear all this, to learn to face the cycles of living and dying, to live within the limits of life, to stop trying to escape into future solutions? Ecologist Paul Shepard (2002) urges us to learn to:

> participate appropriately in the world, to limit ourselves, to acknowledge that we are a part of world food chains (…). It means accepting the world as given, rather than made, a world of limits, contingency, the courteous readiness of the sacramental reality of death. (…) We have scarcely begun to discover what it means to be an organism on a very small planet, from which there is no escape, no alternative. (p. 259).

If we do not learn to do this, we can continue to deny the illness of our bodies and the relationship of that to the death of the whales and so many other species. We can cram 30 children, sweaty and hot, into a small tight space without a window where breathing seems restricted and *think this is fine*. We can unground and un-time ourselves, pretend and even convince ourselves we aren't linked. We can say to children 'sit still' and mean it. We can pretend there are no limits to the growth of populations or economics or consumerism. But then, in the end, the limit is there, and we see this in the extinction of the whale, or in the lack of justice, the suffering of our human neighbour, and in the stripping of the land of the very resources we need to live and breathe, for life to go on.

This is the turning point.

River Walk: Coda

Walking here, life seems lush, flowing, moving, diverse. Year after year the clematis has bloomed. Who would notice if it disappeared? Would it matter? What do they depend upon and what depends on them? Does the intimate cycle of the watershed come to mind at this moment? Where this water comes from and where it goes?

On this day, there is a commotion near one of the pedestrian bridges. A gathering crowd and a police car. A young man, dressed in a dark track suit, is perched out beyond the rails, precarious on the concrete bridge abutment high above the cold water. Will he jump? Someone on the bank is taking a photo of a young police woman crouched down, holding out her hand towards the man, reaching her life towards his. Everyone feels the crisis, the turning. Who among us does not know this feeling of life losing its future possibilities, openness, wonder, this terrifying aloneness that comes with closure and despair?

This time, life takes a breath.

He reaches his shaking hand towards the police woman.

Contact.

I do not know this young man on the bridge, will never see him again, but I walk the rest of the way home with an aching spirit in this close intimate time of another's mortality, in this place that holds the memories of glacier ice, blue clematis and wild sweet roses about to bloom.

14. *In Praise of Radiant Beings*

DAVID W. JARDINE

A Preambling Couplet

"To hold that the world is eternal" the Buddha declared, " … is the jungle of theorizing, the wilderness of theorizing, the tangle of theorizing, the bondage and the shackles of theorizing, attended by illness, distress, perturbation and fever." It is important to assimilate this passage in its entirety. It points to a reality that transcends ordinary thought but is nevertheless still knowable. To say that it is possible to know something that is beyond thought carries the important, indeed astonishing implication, that *there is in the mind a dimension that in the vast majority of living beings is wholly concealed, the existence of which is not even suspected.*

from the translator's introduction to *Introduction to the Middle Way: Chandrakirti's [c. 7th Century CE]* Madhyamakavatara *with Commentary by Jamgon Mipham [1846–1912]* (2002, p. 9, emphasis added).

We will have to hold firmly to the standpoint of finiteness. [This] does not mean that [the human subjectivity] is radically temporal, so that it can no longer be considered as everlasting or eternal but is understandable only in relation to its own time and future. If this were its meaning, it would not be a critique and an overcoming of subjectivism, but an "existentialist" radicalization of it. The … question involved here … is directed precisely at this subjectivism itself. The latter is driven to its furthest point only in order to question it. In disclosing time as the ground hidden from [subjective] self-understanding it … opens itself *to a hitherto concealed experience that transcends thinking from the position of subjectivity.*

from H.G. Gadamer [1900–2002], *Truth and Method* (1989, pp. 99–100, emphasis added).

"Even There"

> *Kai enthautha*, "even here," at the stove, in that ordinary place where every thing and every condition, each deed and thought is intimate and commonplace, "even there" *einai theous*, "the gods themselves are present." (Heidegger 1977b, p. 234).

Spending time in schools where good work is being done has become increasing hard for me to bear, not because of something negative (this, of course, has often taken its own toll) but because of something profoundly positive and nearly unutterable. There is a practice at the heart of hermeneutic work (a practice shared in various and varying ways with ecological awareness and threads of Buddhist philosophy and practice) that results, mostly gradually, but sometimes suddenly and without warning, in the ability to intimately and immediately experience the dependently co-arising (Sanskrit: *pratitya-samutpada*) reality of things, ideas, word, selves, gestures, actions.

With deliberate practice, this ability, "hitherto concealed" by distraction and grasping at straws, builds, and as it builds, so too the building of the world and its ways (Dharma) flutters open and its interdependencies become more and more experienceable (a process parallel to Hans-Georg Gadamer's work [1989, pp. 9–18] on the old German idea of *Bildung*, where "becoming experienced" leads to an *increasing* susceptibility and vulnerability in one's ability to experience of the fabrics and textures of the world), more and more *loveable*, "despite all my revulsions over its ugliness and injustice, and my bitterness over defeat at its hands" (Hillman 2006a, p. 128).

"Even here," "not even suspected," the very tiniest and most meager of things, with experience and practice, can come to be experienced in beautiful repose.

Ah! Wabi Sabi

What is of concern here in this chapter is something extremely simple, but it is a simplicity in which is concealed something miraculous which, when it appears, is often treated too casually, laughed off in ways that do not recognize the deep aesthetic truth that comes when the breath halts and the body leans over in laughter (Jardine 2012b).

Consider: When a young child tips forward into a word and finds herself struggling to sound it out, humming and murmuring over its sonority and the ancient, specific, detailed links lurking there with the look of its letters, its words, its spaces, tossing it around her tongue and breath, tripping unknowingly over all those old mongrel roots of English, all inbred and tangled together and pushing and pulling of attention this way and that, perhaps not yet sensing the haunting presence of such forgotten ancestors and ancestries that are ripe and ready to be known even though we can get along quite famously in sheer ignorance of their life and lives. This simple act of pronunciation is at once the most ordinary of classroom events and a great, roiling thing, a great nexus full of the "silence of a world turning" (Domanski 2002, p. 245).

And then suddenly, yes!

Pronounced.

This word and *that* sit ready to be carried up out of the dust of written letters and outwards on the voice, held up on great pillars of breath, enspirited ("*aesthesis*, which means at root a breathing in ... of the world, the gasp, 'aha,' the uh' of the breath in wonder ... and aesthetic response" [Hillman 2006, p. 36]) and at once conspiratorial (Illich 1998) because here I am, sit squat beside her in breathless anticipation of utterance.

"All writing is a kind of alienated speech, and its signs need to be transformed back into speech. This transformation is the real hermeneutical task" (Gadamer 1989, p. 393).

Ah!

So *this* is what those words speak. *Wabi Sabi* (Reibstein & Young 2008), that wandering cat off to find what her name means through the scrolling pages of this picture book. That young child's finger now pointing to the great, collaged images on the page, and then back to the cover, her hand rubbing its rough smoothness, and then, with the sound and the sight of this name now lodged in memory, safe for now in their keep, the story then continues to unfold, her smiling and pointing, me nodding in agreement, after all that tough and fruitful work along this book's path. And this young child's eye has ripened in its ability to scan the clusters of scrolling text for this cat's name's next appearance.

I have seen this so many times, such small events that have never happened before, here, now. "So that ... sitting there, listening like that, becomes part of the story too" (Wallace 1987, p. 47), me, a reader, having learned to read, having read countlessly about reading and its varying courses of ancient emergence, having been in hundreds of classrooms and felt this halt of breath over and over and over again, the life-breath of sounding out, as if the whole of the language itself depends for its very continuance, its very life, on this next moment's precarious venture. Again, and again, "where it seems impossible that one life even matters" (Wallace 1987, p. 111), a small, simple, innocuous event that could easily be deemed perfectly trivial in the grand scheme of things becomes experienced as sitting in the center of worlds of relations, a residence, a "housing" (*ecos*). Right there in the very ordinariness and mundaneness of this event ("right now … *this*" [Wallace 1987, p. 111],) we become huddled near the living origin of language itself, the very moment of its (oddly old-yet-brand-new, all at once) emergence, sustenance and survival. In this way, as a teacher, and in this smallest of examples, I get to be present as the world of language is being "set right anew" (Arendt 1969, p. 197), right before my eyes and ears, revived, saved from its mortal lot just in the nick of time.

This sort of experience has become, for me, a simply *miraculous* thing to be around, something almost unbearable in its countenance. What is most deeply experienced here is the lovely, hearty frailness of these strange human ventures of reading, of speaking, of voice and utterance, and how the ancients in all human traditions have gathered around such ventures, most often unheralded and forgotten, and in myriad ways. They, too, are right "even here" in this act of breath, this pronunciation, holding their breath and ours in anticipation.

Hermeneutics and threads of the Gelug tradition of Tibetan Buddhism provide, in very different guises and for similar but still different motives, ways to articulate this kind of conspiratorial and deeply pedagogical experience of the world. Both detail how this way of experiencing the world can be cultivated through long and difficult practice ("one must learn how" [Gadamer 2007b, p. 217]). They also share something of an *ontological theory* that attests to the reality of these experiences of dependent co-arising.

Both bespeak a "wisdom [which] thoroughly discerns *the ontological status of the object under analysis*" (Tsong-kha-pa 2004, p. 211). The tough work of these scholarly traditions and practices helps keep at bay the commonplace trivializing of these experiences. The interlacing kinships between Gadamerian hermeneutics and the Gelug lineage of Tibetan Buddhism provide ways to elaborate what has become, for me, an elusive yet familiar experience in the haunts of schools, ways to remain with and true to these experiences.

"Protodoxa (Urdoxa)"

> To realize the full import of dependent-arising, namely that all phenomena are empty of inherent existence, is an extremely forceful experience that reorients one in the very depths of one's being. (Lobsang 2006, p. 51).

> By virtue of repeated practice, you become free of your dysfunctional tendencies, undergoing a fundamental transformation. (Tsong-kha-pa 2002, p. 36).

> Indeed, this true mode must include … a conversion of the standpoint of Reason. (Nishitani 1982, p. 117).

> Perhaps it will become manifest that the total phenomenological attitude [is] destined in essence to effect, at first, a complete personal transformation, comparable in the beginning to a religious conversion. (Husserl 1970, p. 137).
>
> An *inner transformation*. (Husserl 1970, p. 100).

One element of Hans-Georg Gadamer's philosophical hermeneutics is an *ontological insight* into how things, ideas, images, and selves *exist*, their manner of Being. This insight is in lineage back through the work of Martin Heidegger and is rooted in the work of their teacher, Edmund Husserl, the "father" of contemporary phenomenology.

This ontological insight involves a *critique of substance*—an age-old idea that runs back through the work of Descartes (1955, p. 255) in the 1600s and is rooted in Aristotelian ontology: "A substance is that which requires nothing except itself in order to exist." Under such an auspice, any thing, image, idea, concept, object, or self *is what it is independently of anything else. To be* something real is *to be* separate, substantive, and independent, a "permanent, unitary and autonomous entit[y]" (Yangsi 2003, p. 241), something thus "inherently self-existing" (Tsong-kha-pa 2002, p. 120). To understand anything in the world (like that young child's efforts at pronouncing that new name found in a new book) is to separate it off from everything else. We must adopt a "reifying view" (Tsong-kha-pa 2002, p. 120) regarding the ontological status of the thing being experienced: whatever this thing is, *it is what it is*, and it is not something else. Simple. We may not know *what* it is, *why* it is thus or *how it came to be* thus, but we know, with this reifying view, *that* it is what it is. So if we have no specific knowledge of it at all and it is simply some unknown "X," we *do* know that, whatever this "X" turns out to be, "X = X."

To use the language of Edmund Husserl's phenomenology (1969, p. 106), we may have doubts about this thing but, despite all this trepidation, "the 'it' remains ever in the sense of a general thesis, a world that has its being 'out there'." Husserl (p. 169) called this the "general thesis of the natural attitude." It is, as he suggested, not simply one belief among others, but *the* founding belief of the commonplace way in which, "in everyday life our minds apprehend [things] as existing" (Yangsi 2003, p. 200). Husserl (1969, p. 300) named this "the *primary belief* (*Urglaube*) or *Protodoxa* (*Urdoxa*)" of the natural attitude. Note, however, that *in* the natural attitude, this Urdoxa is understood to be precisely *not* "a belief" but simply "the way things are." It is assumed and projected as an ontological "given" against which our everyday experience of the world is to be understood. This is the odd breakthrough of Husserlian phenomenology, unearthing this Urdoxa as a thesis *of* the natural attitude that does not appear *as* a thesis *in* our ordinary experience of the world. This thesis thus describes the founding yet concealed and unquestioned prejudice of everyday life.

Within this (as Buddhism would have it, false or delusional) *belief*, "the essence of truth is identity"(Heidegger 1978, p. 39), and the formal consort of this presumption of identity is the mathematical and formal-logic principle of identity (X = X). This thus links up the possibility of *understanding* the substance/reality of things with logico-mathematically based, natural-scientific methodologies, concepts and categories. Since the thing itself is what it is (X = X), *knowledge of that thing* must itself have precisely such clarity and distinctness borne of one thing separated off from another.

"Break Open the Being of the Object"

> Because this causes living beings to be confused in their view of the actual state of things, it is a delusion; ignorance mistakenly superimposes upon things an essence that they do not have. It is constituted so as to block perception of their nature. It is a concealer. (Tsong-kha-pa 2002, p. 208).

The hermeneutic critique of the dominance of natural-scientific methodologies sits squarely here, on a critique of the idea of substance—this concealed Urdoxic belief in separate and inherent self-existence (Sanskrit: *svabhava*) as a formulation for how things exist. Gadamer (1989, p. 242) states this directly: "The concept of substance is ... inadequate for historical being and knowledge. [There is a] radical challenge to thought implicit in this inadequacy." Understood hermeneutically, interpretation and questioning are therefore not just a matter of "making connections" between two inherently separate things (we are not dealing here with the *epistemology* of constructivism, where a subject puts separate things together and thus "produces" connections). Rather, interpretation and questioning have an *ontological* force: they "break open the [falsely presumed to be self-existent and substantive] *being* of the object" (Gadamer 1989, p. 362), making visible, experiencable and understandable the ontological inherence of one thing in the very being of another.

In Buddhism, this phenomenon of "cosmic interpenetration" (Loy 1993, p. 481) is described in conjunction with another equally important ontological insight: emptiness (Sanskrit: *shunya*). This insight is named and/or translated in various ways: things, ideas, images, selves, are considered to be "empty of self-existence (*svabhavasunya*)" (Tsong-kha-pa 2000, p. 24), "empty of having an inherent self-nature" (Tsong-kha-pa 2005, p. 183), having an "absence of self-nature" (Tsong-kha-pa 2000, p. 20), possessing "not even an atom of ... true existence" (Tsong-kha-pa 2004, p. 215):

> The true mode of being of a thing as it is in itself, is selfness, for its self cannot be a self-identity in the sense of a substance. Indeed, this true mode must include a complete negation of such self-identity. (Nishitani 1982, p. 117).

However, this "negation of self-identity [X = X]" does not lead to something null and void. "Empty of inherent existence" (Tsong-kha-pa 2006, p. 33) is meant to point towards *how* things *do* exist—not as separate self-identical substances that need nothing except themselves to exist, but as, rather, "dependently co-arising" (Sanskrit: *pratitya-samutpada*): "The only way phenomena do exist is as interdependently related" (Lobsang 2006, p. 51). Emptiness of substantive self-existence thus is identical to the fullness of dependently-arising interrelatedness that defines things, words, objects, and selves. Things *are* thus—in the very being of every seemingly separate thing are nestled *worlds of relations* and our ordinary experience of this as simply a meager act of pronunciation leaves such worlds concealed. This is what is meant by the breaking open of the (falsely presumed to be self-enclosed) being of the object and thus releasing insight into the instead dependently co-arising being of the object. "Dependent-arising is the meaning of emptiness" (Tsong-kha-pa 2002, p. 133). Emptiness (of separate self-existence) thus means fullness (of dependently arising existence).

"Every Word Breaks Forth"

There is a passage I have always loved in Hans-Georg Gadamer's *Truth and Method* (1989, p. 458) where he portrays something of the experience that follows from this gathering sense of "break[ing] open the being of the object":

> Every word [has an] inner dimension of multiplication: every word breaks forth as if from a center and is related to a whole, through which alone it is a word. Every word causes the whole of the language to which it belongs to resonate and the whole world-view that underlies it to appear. Thus every word, as the event of a moment, carries with it the unsaid, to which it is related by responding and summoning.

This "*as if* from a center" is vitally important.

As if, because "*the center is everywhere.* Each and every thing ["every word"] becomes the center of all things and, in that sense, becomes an absolute center. This is the absolute uniqueness of things, their reality" (Nishitani 1982, p. 146). "We should apply this [as if] to *every* phenomenon. Every phenomenon ... is empty of having an inherent self-nature that exists from its own side" (Tsong-kha-pa 2005, p. 182). So that at "the periphery" of experiencing that young child's slow and agonizing work of pronunciation resides, for example, the phenomenon of names and how we are called by them, how they "summon and respond," or the appearance of spaces between words into written English in the 11th century (see Carruthers 2003, Illich 1993, Illich & Sanders 1988, Stock 1983). There, too, is the example of a colleague (see

Jardine & Naqvi 2012) telling me of her daughter learning to read the Koran out loud, and how she often did this well-honed pronunciation without understanding what many of the words mean. I queried this and found that *uttering the very sound of the words themselves* was understood to be a mirroring of the very sound of God speaking to the prophet, so the very sonority itself was blessed and precious, independently of the meaning and message of the text. Yes, oral recitation, a commonplace of Canadian elementary school classrooms. Reading aloud, "even there." God uttering the world into existence over the face of the deep. And that silent reading didn't even enter European consciousness until around the 11th century and, in doing so, propagated a new understanding of ourselves and our "interiority" and individuality/privacy (see Illich 1993, Illich & Sanders 1998). And there, too, is the full cascade of cultures and tongues that have found their ways here into this classroom, and how each has suffered in their own way the great modern hegemony of the English tongue (even, of course, for native English speakers), even here, in efforts to pronounce this Japanese transliteration of a deeply culturally embedded Japanese name for a Japanese cat in this child's new book.

So that slow breath of pronunciation is thus experienced, not as an isolated object of consideration, but as existing in a broad and generous "residence" of possibilities, lineages, intergenerational bloodlines, and the like. Hermeneutically understood, every however-common event of language *is* a dependently co-arising, living inheritance, and therefore the contingent, localized, frail and fragile *taking up of that inheritance*—this girl's efforts, here, now—is (however small a) part of its being what it is. As is my own inhaled breath, here, in this classroom, witnessing all over again the arrival of this old, familiar, tough-minded companion: pronunciation.

This *seems* paradoxical, that the center is everywhere and that, therefore, "none is the fundamental entity" (Hahn 1986, p. 70) while, at the same time, any thing can be experienced *as* ("if") the center of all things. This is only a "contradiction" against the presumed backdrop of the logic of substance. Keiji Nishitani (1982, p. 149) elaborates:

> To say that *a thing is not itself* means that, while continuing to be itself, it is in the home-ground of everything else. Figuratively speaking, its roots reach across into the ground of all other things and help to hold them up and keep them standing. It serves as a constitutive element of their being. *That a thing is itself* means that all other things, while continuing to be themselves, are in the home-ground of that thing. This way that everything has being on the home-ground of everything else, without ceasing to be on its own home-ground, means that the being of each thing is held up, kept standing, and made to be what it is by means of the being of all other things; or, put the other way around, that each being holds up the being of every other thing, keeps it standing and makes it what it is.

Our considerations can thus then shift to one of these constitutively surrounding things that constitute the peripheries of pronunciation. Such shifts would then make her efforts of pronunciation now peripheral to, but still constitutive of this new "as if" center of our attention. *As if*, because anything "at the center" *is* "itself" because it *is* the field of its residing.

Thus Gadamer's "*as if* from a center," because in looking "into" that center and focusing in with great care on that child's pronunciation, enjoying, praising, encouraging, waiting, laughing, demonstrating, we don't experience it like some sort of some hard-shelled "core." In looking into that center, we are cast "outwards" into worlds of relations. We are drawn "into" this child's efforts and find that in being drawn into that center, it "draws us entirely outside of ourselves. Rather than meeting us in our world, it is much more a world into which we ourselves are drawn. *[T]he totality of a lived context has entered into and is present in the thing*. And we belong to it as well" (Gadamer 1994, pp. 191–2, my emphasis). This is why, in nearing such events in a classroom, my own life as a teacher *necessarily* involves learning, because in properly taking up pronunciation as a teacher, I take up, inevitably, the implication of *my own being* in this phenomenon *and* my dependently co-arising responsibility for the well-being of this girl and this inheritance and our myriad places in this great residence.

This is why I am drawn to such events in the classroom, "responding" to this "summons," because, hermeneutically understood, I, too, am already "present in the thing." In interpretive work, then, we come to "recognize *[our] selves* in the mess of th[is] world" (Hillman 1983, p. 49, emphasis added), by recognizing ourselves, not as some self-existent entity, but rather *as* the mess of the world. I myself, as with this young girl, as with her breath and mine—each of these is experienced as glancing through the reflected facets of the world, each being itself by being such glancing with no substance left over.

A metaphor for such cosmic interpenetration and lack of self-presence is found in the Avatamsaka Sutra of Mahayana Buddhism: Indra's Net:

> Far away in the heavenly abode of the great god Indra, there is a wonderful net that has been hung by some cunning artificer in such a manner that it stretches out infinitely in all directions. In accordance with the extravagant tastes of deities, the artificer has hung a single glittering jewel in each "eye" of the net, and since the net itself is infinite in all dimensions, the jewels are infinite in number. There hang the jewels, glittering like stars of the first magnitude, a wonderful sight to behold. If we now arbitrarily select one of these jewels for inspection and look closely at it, we will discover that in its polished surface there are reflected all the other jewels in the net, infinite in number. Not only that, but each of the jewels reflected in this one jewel is also reflecting all the other jewels, so that there is an infinite reflecting process occurring. (Loy 1993, p. 481).

This not only means that we "are always open onto the horizons of others but also, more important, because [we] are *always already everywhere inhabited* by the Other in the context of the fully real." (Smith 2006, p. xxiv). Oddly put, experiencing that young girl's efforts at pronunciation requires my experiencing that, in my fullest reality as dependently co-arising, *I am*, my very self, a dependently co-arising inhabitant of this very residence.

I am inhabited by her halting breath over that odd name. It is my very life being played out, here, in watching her paw that book's cover and loving its allure, just as surely as my life is played out in the grief I sometimes find in schools. Coming to understand, then, is "more a passion than an action" (Gadamer 1989, p. 366), more an act of compassion for the suffering and impermanence that defines our deeply shared, often deeply concealed lot.

This is my own column of aging breath being summoned up in this event of pronunciation. In breaking open the being of the object, *my own being "myself" is broken open* and experienced as caught up and constitutively implicated in the very fabric of the object I meditate upon or study or happen upon in some local classroom. This great sentiment is so eloquently expressed by Rick Fields (cited in Ingram 1990, p. xiv):

> My heart is broken,
> open.

This is why it is getting so tough to visit schools sometimes. That child's pronunciation venture is heartbreaking.

Therefore, in "[heart]breaking open the being" of this child's efforts at the pronunciation of a name, "we are not attempting to get rid of [it], only of the idea of [it] as self-existent" (Lobsang 2006, p. 49). "Those objects that appear ... do not stop appearing, but the concepts [e.g. "substance," or other reifications] that take them as having any true existence subside" (Patrul 1998, p. 252). We still remain concerned after understanding pronunciation in the specificities of its appearance, here, with this child and this book and this name and all the oddities of its translations and transliterations. But, at the same time, "the object" of our consideration must also include "the emptiness of this [particular phenomenon], not just [the phenomenon]." (Yangsi 2003, p. 433)

To use the language of Martin Heidegger (1962), our concern is not merely "ontic" ("What is this thing, pronunciation, and what are the threads of its dependent co-arising?") but "ontological" (our "ontic," interpretive concern for this particular entity is only authentically pursuable against the background of its *being* dependently co-arising). From Longchenpa (1308–1363):

> Knowledge is as infinite as the stars in the sky;
> There is no end to all the subjects one could study

> It is better to grasp straight away their very essence—
> The unchanging fortress of the dharmakaya. (cited in Patrul 1998, p. 261).

("Dharmakaya—*chos sku*, lit. Dharma Body. The emptiness aspect of Buddhahood" [a glossary entry to Patrul 1998, p. 410]). Thus "studying subjects" like pronunciation must be done against the ontological "backdrop" of a knowledge of emptiness (a knowledge, that is, that pronunciation is *interpretable*) and these two ways of proceeding support and cultivate each other. Careful attention to the great detail, specificity and particularity of *this* appearance of pronunciation ("distinguishing the exact particulars of an object" (Tsong-kha-pa 2002, p. 17) is possible *because* pronunciation is treated, interpretively, as empty of self-existence, as breaking forth *as if* from a center. And likewise, in interpreting pronunciation, we can get a glimpse of, an experience of, emptiness.

"Accordingly there is the examination and analysis of *both* the real nature [emptiness—the 'ontological status of the object under analysis' [Tsong-kha-pa 2004, p. 21] and the [ontic] diversity of phenomena" (Tsong-kha-pa 2002, p. 17). Properly understood, between these "there is compatibility and a lack of contradiction" (Tsong-kha-pa 2004, p. 215). (This helps unravel a bit why Martin Heidegger always insisted that the Being of beings is not a being *and* he insisted that Being is always the Being of a being—I'll leave the tempt of this tangle for others to undo at their leisure).

"It Draws You into Its Path"

> When [it] takes hold of us, it is not an object that stands opposite us which we look at in hope of seeing through it to an intended conceptual meaning. Just the reverse. The work [e.g. of pronunciation] is an *Ereignis*—an event that 'appropriates us' into itself. It jolts us, it knocks us over, and sets up a world of its own, into which we are drawn, as it were. (Gadamer with Dutt 2001, p. 71).

There is, of course, a danger here. The hermeneutic experience of the breaking open of the being of the object involves "the way you apprehend the object. By making the object extensive [you] expand your mind" (Tsong-kha-pa 2002, p. 63). As "it" becomes more extensive we become, as St. Augustine put it, "roomier"(cited in Carruther's 2005, p. 199). My "self" expands as my apprehension of the dependent co-arising of the world—the "residence" of my self—expands: "Making the object of meditation extensive so as to expand your mind" (Tsong-kha-pa 2002, p. 63).

And as the object "breaks forth," "it draws you into its path" (Gadamer 2007b, p. 198).

Hence arises the danger that many new to hermeneutics (and Buddhist meditative practices) often experience: *Everything* seems connected, *everything* is rampantly full, *everything* starts to beckon and summon. It expands and I expand and get caught up in an onrush that seems impossible to stop. This is sometimes named the "monkey mind" in Buddhist practice (and this breaching an ironic humiliation in the *Wabi Sabi* book, where Monkey is the teacher).

> The more you practice these things, the more accustomed your mind will become to them, and the easier it will be to practice what you had initially found difficult to learn. You will have visions of the Buddha day and night. (Tsong-kha-pa 2000, pp. 185–6).

> One arises from formal meditation and goes about daily activities, seeing the manifestations of the world and living beings as mandala deities. This is the Samadhi that transforms the world and its living brings into a most extraordinary vision. *All* experiences ["even here"] are taken as manifestations of great ecstasy. (Tsong-kha-pa 2005, p. 125, emphasis added).

Everything is connected to *everything*. And hence a common complaint for some students of hermeneutics: "How do I get it to stop?" or "*Now* what?"

This outward expansiveness can easily lead to a ravaging exhaustion of attention, a sort of cascading, post-modern "connectionism": "You [can easily become] like the leading edge of water running downhill—you go anywhere you are led, taking anything said to be true, wanting to cry when you see others crying, wanting to laugh when you see others laugh" (Tsong-kha-pa 2004, p. 222). It is easy to become swamped by possibilities and simply outrun. This is why Gadamer (1989, p. 106) talks about how it is that the dependent co-arising of play of things (e.g., what is expandingly experienced as "at play" in that breath of pronunciation) can "outplay the players."In the face of that simple event in an elementary school classroom, in the face of that lovely book and its allure, one can become simply burdened by unseemly and unending cloys of "relatedness" which can weigh down attention and make me simply give up in overwhelmed frustration. Tsong-kha-pa (2002, p. 62) thus warns against the extremes of "laxity [which simply gives up, spent in the face of the rush] and excitement [which simply pursues the rush ever faster and with accelerating distraction]."

A key task of hermeneutics as a practice (as well as a task in Buddhist practice) is thus the practice of *composure* in the face of this ecstatic experience of breaking forth. Tsong-kha-pa insists that the insight into emptiness that comes from pursuing wisdom be coupled with the practice of meditation, of one-pointed stillness and composure. He uses the term "equanimity" (2002, p. 68) to describe a process of "relaxing the effort, but not sacrificing the intensity of the way you apprehend the object" (p. 68). What is called for, then, is facing the delusions of inherent self-existence and *interpreting* them, that is, breaking open these delusions so that the realities of dependent co-arising can be experienced and understood. But this must be taken on at the same time as *not* simply becoming caught up in mere pursuit of the ensuing cascades. "I compose this in order to condition my own mind" (Tsong-kha-pa 2000, p. 111).

Moreover, I read these compositions of Gadamer and Tsong-kha-pa in the very same pursuit of conditioning my own mind. "Texts are instructions for [the] practice" (Tsong-kha-pa 2000, p. 52) of precisely paying more proper attention to that girl's way through the sounds of words.

It is notable, then, that we are right back to the issue of proper pronunciation and of learning to read, and, right here, in the face of this ordinary classroom event, this is, of course, exactly where we should be. Admittedly, it is difficult to read many scholarly texts this way, as instructions for practice, and this, too, takes practice, just *exactly* as does learning to "read" that singular moment in the classroom as "breaking forth," that is, as an exuberant "event of appearing [*Vollzug*]." About such moments, we ask:

> [how does] it begin, end, how long [does] it last[?]; how [does] it remain in one's mind, and in the end how [does] it fade away, and yet somehow remain with us and [be] able to surface again[?] (Gadamer 2007b, p. 217).

And this faces us with the dilemma: what should I do now? What should I say or write or show or save? What should be highlighted and what forgotten? This takes time, but a certain kind of time that is itself meditatively proper to such composition:

> Certainly one can call this process a "while" [*Weilen*], but this is something that nobody measures and that one does not find to be either boring or merely entertaining. The name I have for the way in which this event happens is "reading." With reading one does not imagine ... that one can already do it. In reality, one must learn how .... Now the word *Lesen* ["read," a German kin to the English word "lesson"—I think, for example, of an old commonplace in Anglican church services, of saying "today's lesson is taken from Matthew," meaning both literally "a reading from Matthew" but also *reading* that reading for its "lesson"] carries within it a helpful multiplicity of harmonic words, such as gathering together [*Zusammenlesen*], picking up [*Auflesen*], picking out [*Auslesen*], or to sort out [*verlesen*]. All of these are associated with "harvest" (*Lese*), that is to say, the harvest of grapes, which persist in the harvest. The word *Lesen* also refers to something that begins with spelling out words, if one learns to write and read, and again we find numerous echo words. One can start to read a book [*anlesen*] or finish up reading it [*auslesen*], one can read further in it [*weiterlesen*], or just check into it [*nachlesen*], or one can read it aloud [*vorlesen*]. All of these point towards the harvest that is gathered in and from which one takes nourishment. (Gadamer 2007b, pp. 217–8).

Harvest, gathering, nourishment. Lovely words to describe the yield that comes from the suffering undergone in learning to read. To pronounce.

This is why Gadamer follows his lovely exploration of the phenomenon of the play(s) of the world (1989, pp. 101–110)—like being drawn into what is "at play" in that girl's work with *Wabi Sabi*—with an imposingly title section "Transformation into Structure" (pp. 110–120). "Transformation into structure" is about creating, in the midst of the cascading and alluring play of the world that is experienced once it is broken open, a "play," a "work" or a "composition" which draws us into "what is at play" and makes possible a certain composure in the face of this insight. "Composition" thus "captures" something of the insight of breaking forth and, with practice ("one does not imagine ... that one can already do it. In reality, one must learn how" [Gadamer 2007b, p. 217]) a work can be shaped that can offer to others the possibility of engaging this insight. This is identical to the everyday work of a classroom that gathers students together to create beautiful things that will show others how their work is laced up in the dependent co-arising work of the world. In taking on the difficult work of composing such events of appearing, one offers an invitation into such arising worlds. They are thus things that allow me, invite me, into a "hitherto concealed experience that transcends thinking from the position of subjectivity" (Gadamer 1989, p. 100), an experience of my own dependent co-arising in the face of such arisings.

This act of, shall we say, mutual composure, then, is not just a matter of achieving stillness in the face of the fray, but of cultivating a deep and well-studied knowledge of the topic of one's meditations and all the rich lineages and ancestries of thought that have handed this topic over to us and us over to it. It involves realizing that "you can't get anywhere without [also] reading a yak's load of books" (Tsong-kha-pa 2004, p. 219). This study is a study *on behalf of* the practice of composure in the midst of every life. It is a matter of not simply slavishly following the heavy scholarly weigh of "tradition" but of, odd to say, having "delighted those who have come before" (Dudjom & Dorje 2011, p. 16) as a way to thus oneself experience delight in this new arising. This is why Gadamer (see Grondin 2003, p. 333) expressed soft and generous disappointment when he found certain works given in his honor at his 100th birthday celebration not very *lebensweltlich* (not very "life-worldly"). This accounts, also, for why, as a hermeneutic philosopher, I feel such an affinity to the Gelug stream of Buddhism that emphasizes the equanimity between study and practice. "Wisdom and the study that it causes are indispensable for proper practice" (Tsong-kha-pa 2004, p. 220).

"Do not make study and practice into separate things" (Tsong-kha-pa 2004, p. 221).

"Against fixity ..." hermeneutic work, strangely akin to Buddhist thought, "makes the object and all its possibilities fluid" (Gadamer 1989, p. 367). Therefore, the purpose of a "hermeneutic study"—the purpose of composure, of composition—is to create something that will draw others into this hermeneutic experience of fluidity—*into* an experience of the dependent co-arising of the world and *out of* the delusions of substance and the grasping-at-permanence syndrome that is its root. And all this has as its aim not simply philosophical erudition and the like. The aim of such meditations and work is the cultivation of the intimacy and immediacy of the experience of everyday life, here, as this next child draws breath over a text, here, where reading aloud and learning to pronounce can too often be treated as simply ordinary and commonplace. Not only is "wherever you are ... a place of practice" (Tsong-kha-pa 2004, p. 191). Tsong-kha-pa also insists (and this is a feature that distinguishes the Gelug tradition from other Buddhist lineages and makes its affinity to hermeneutic ripe), the purpose and object of scholarly study is *precisely* the deepening of practice itself. After all, "why would you determine one thing by means of study and reflection, and then, when you go to practice, practice something else?" (2000, p. 52). And this is why Gadamer insists that hermeneutics, with all its philosophical erudition and study, is "a practical philosophy" (2007c) with both theoretical and practical tasks (2007d). All those complex philosophical and historical twists and turns that typify his work are meant, in the end, to make us more susceptible to the beautiful abundance of things as we walk around in the world. "Even here," in this ordinary place.

"Present in the Thing"

> [In learning to experience events in this way], we experience an absolute opposition to [our] will-to-control ["the syndrome of grasping at a self-nature" (Tsong-kha-pa 2005, p. 182)], not in the sense of a rigid resistance to the presumption of our will which is bent on utilizing things, but in the sense of [having come upon] the superior and intrusive power of a being reposing in itself. (Gadamer 1977, pp. 226–7).

All those complex lines of dependent co-arising that break forth from this event of pronunciation are experienced as being "present in the thing" (Gadamer 1994, p. 192). This is not exactly a matter of attention "narrowing" back in on some separate thing, but rather of attention *intensifying* and this wee event becoming, one might say, "luminous," radiantly full of all its relations. This is akin to accounts of what is called "generation stage meditation" in the Gelug tradition of Buddhism:

> You develop the ability to see yourself as a ... deity and your environment as the mandala or ... abode of the deity. This is called *pure perception*: you perceive all objects of the physical senses and the mind the same way a meditational deity experiences them [as if sitting at the center of an open field of dependent co-arising, a mandala-like "residence"]. The opposite of this mode of perception is *impure perception*. This is ordinary appearance: seeing yourself as ordinary, seeing the world where you live as ordinary ["each deed and thought ... intimate and commonplace"], and experiencing ... objects ... as ordinary ["familiar"]. This impure appearance is the main target to be removed by ... practice. (Sopa 2004, p. 30).

Tsong-kha-pa elaborates (2005, p. 182) that what is sought is the ability to see every phenomenon as such a radiant deity:

> We should apply this to *every* phenomenon. Every phenomenon ... is empty of having an inherent self-nature that exists from its own side. One should understand exactly how strong the presence of the syndrome of grasping at a self-nature [X = X] is. On the basis of [applying this to every phenomenon] one transforms the ... world and its inhabitants [from being experienced as separate, controllable, graspable substances] into the supporting and the supported mandala (i.e., residence and deities).

Picturing every phenomenon as a deity sitting in the middle of a mandala entails coming to experience every phenomenon as thus a *radiant being*, illuminatingly full of all its relations precisely by being empty of self-existence, a being in whose presence light is cast upon me, illuminating my own culpabilities. Such picturing, which arises as an often-agonizing of sustained practice:

> It is not merely one's "taking time" to linger over something, as in the slackening or slowing down to contemplate. [It is] not a function of lackadaisical, meandering contemplation, least of all passive in any way, but is a function of the fullness and *intensity* of attention and engrossment (Ross 2006, p. 109).

In cultivating this "hermeneutic experience," those radiating lines of dependently co-arising venture can be sensed shimmering all around "every phenomenon," akin to a "halo" of illumination. Jigme Lingpa (1729–1798): "Concentration is to experience as deities all the appearances to which one clings" (cited in Patrul 1998, p. 254), that is, as a footnote to this passage clarifies, "to meditate on appearances as deities ... means as pure wisdom manifestations [pure manifestations of an emptiness of substantive self-existence] with no concrete reality" (Patrul 1998, p. 389, fnt. 166). With repeated and intense practice, even an ordinary appearance like the breath of pronunciation, shall we say, "glows."

That is, this girl's pronunciation is *beautiful* and, through practice, my own life can become more open and illuminated in its presence:

> That which manifests itself [in this way] attracts the longing of love to it. The beautiful disposes people in its favor immediately. The beautiful ... has its own radiance. For "beauty alone has this quality: that it is what is most radiant (*ekphanestaton*) and lovely" [*Phaedrus*, 250 d 7 (Plato 2012a)]. (Gadamer 1989, p. 481).

That "every phenomenon" (Tsong-kha-pa 2005, p. 182) ("every word") hides this lesson is the great and difficult pedagogical affinity between hermeneutics and this old lineage of Buddhist thought and practice. *This* is "hermeneutic experience," that "the more ... the work opens itself, the more luminous becomes [its] uniqueness" (Heidegger 1977a, pp. 182–3), just this, just here, just now, this lovely, simple, beautiful thing, right in the midst of all this frail and fraught suffering and endurance.

And so we look at each other, and the story continues, and she giggles as she sees the name again.

Post-Ambulatory Couplet

> Knowledge is not a projection in the sense of a plan, the extrapolation of the aims of the will, an ordering of things according to the wishes, prejudices, or promptings of the powerful; rather, it remains something adapted to the object, a *mensuratio ad rem*. Yet this thing is not a *factum brutum* [something "inherently existing "] but itself ultimately has the same mode of being as Dasein [human being]. [*Both* are dependently co-arising]. The important thing, however, is to understand this oft-repeated statement correctly. It does not mean simply that there is a "homogeneity" between the knower and the known. The coordination of all knowing activity with what is known is not based on *the fact that* they have the same mode of being but draws its significance from the *particular nature* of the mode of being that is common to them. It consists in the fact that neither the knower nor the known is [substantively] "present-at-hand." (Gadamer 1989, p. 261).

> Because the ultimate mode of the existence of enlightened [radiant] beings and ordinary sentient beings is the same [even "enlightened beings" are dependently co-arising], we have the potential to be recipients of their divine activity. (Yangsi 2003, p. 133).

15. Some Thoughts on Teaching as Contemplative Practice

JACKIE SEIDEL

> One must not build fantasies around the future and just use that as one's impetus and source of encouragement, but one should try to get the real feeling of the present moment.
>
> —Chogyam Trungpa

Elementary School Truths: Everything and everyone is rushing, rushing, rushing. So many meetings. So much curriculum to cover. So many children with so many needs. And then report cards and playground supervision and teacher workshops. How and where is there time to go to the washroom and eat some lunch? Never mind finding time to contemplate . . .

How are we in education experiencing what Chogyam Trungpa (1991) named this "real feeling of the present moment" (p. 151)? The times seem strained. Perhaps we can feel this in the ways that schools are being squeezed by the pressures and processes of global forces, both economic and political. As teachers, we are told repeatedly by governments and the media that we are not succeeding, that children are not measuring up. Education has been increasingly described around goals of preparing children for work—in and for the global economy—and the word *democracy* slips away, to be replaced by the language of competition, measurement, and bottom lines. In this linguistic slippage, the possibilities for our work with children in schools seem to narrow. Among my colleagues and within myself, I sense a growing fear: Our work will never be enough, our own embodied and experiential wisdom is disregarded, and there will always be another project, another method, another wave of inadequacy and another reform program sold to schools. We feel less freedom to make decisions about children's learning and our teaching. And the children are being measured by test after test that tell us nothing about them as human beings. *We feel so tired.*

Here in the West, we place much faith in the future. Political rhetoric all over North America names children as future workers, as global competitors in a tight market. In such a view, the future is infinitely delayed to a mythical place that never arrives. In such a view, children (and all humans, other species, life) can be seen as resources (or impediments to progress), and the living spaces of life, earth and sea, have become expendable in the names of profit, progress, and nation. Does the line of time draw tight now, strangling us in its narrow hold? Does it cast children out of meaningful time, embodied time and place, and how do we materialize children and ourselves in a living present? *Are we losing our bearings?*

Can we say, with Pema Chodron (1997), that "we live in difficult times" (p. 146)? Vandana Shiva (1997) described the ways that we are in the midst of a crisis of diversity—that is, a profound closing of the future through the simultaneous and related extinctions of language, culture, and global biodiversity, brought about not by natural process but under the legacy of colonialism, industrialization, and economic globalization. What we feel in our classrooms is happening everywhere. The earth is being squeezed and life ground under. *We sense this happening.*

Then where is hope? Can we find it in difficult times? Might it arrive, subtly, softly, through learning and remembering to live and be in the world and in our classrooms in different ways? Perhaps, in the face of these processes of empire, new urgencies and possibilities are emerging for teachers to be among those who take up the most profound questions of human life with courage and dignity. Bringing about in our classrooms a downsizing, toward more silence and fewer words. Showing how we might learn to find wealth and fullness in our relationships, in community, in dialogue. Unfolding toward more slowness, mindfulness, and heartfulness. Or will we, as Emilia Ferreiro (2003) warned, "let ourselves get carried away by the vortex of competitiveness and profitability" (p. 56)? Will we allow our distractedness, our busyness, to consume not just us, but the whole earth, all creatures, all life? The stakes are high. *This is not a test.*

I return, this year [2005], to public school teaching after 8 years of graduate study and university teaching. Committing myself to exploring the meaning of contemplation as part of teaching practices, I am trying to understand the ways that bringing mindfulness to the moment of teaching might be a healing project. I do not want to fall back into old habits and patterns. It is easier for school to be a rushing time, to speed swiftly through each day, to hurry children in their learning. But I vow to learn to go more slowly, to breathe more, to make more space for children's minds and lives and experiences.

August 31, 2005

Tomorrow is the first day of school. The leaves are already turning yellow, red, brown. The air snaps crisp in the mornings. Tomorrow the children arrive. They arrive already labeled and categorized. English as a second language. Refugee. Autistic. Down Syndrome. Attention deficit disorder. Gifted. Behavior problem. Grade 3. Grade 4. Normal. The children arrive to me as categories, even before I meet them or know their names. My body prickles with anxiety. I am not ready.

Breathe.

I meditate on these categories, on the meaning of the labeling and medicating of children so that they can "fit" into the school system. Consciously rejecting this vision, I push away the categories in my mind. Special needs. No. ESL. No. ADHD. No. Down syndrome. No. Gifted. No.

No.

I prepare this room for children to come. I try to make a sanctuary without rigid categories or expectations. Places to be quiet and places to be together. Places for our minds and hearts to meet, to think new thoughts together that we have not thought before. Who knows what can happen? Acknowledging my anxiety and hope and anticipation all at once, I tell myself to slow down. Slow down and breathe.

Yes.

Come in. Welcome.

Bringing mindfulness to the moment of teaching is one way of bringing a deeper timebound and earthbound thoughtfulness to our work with children. It is not easy to enter this mindful space-time in educational places, not only because of our constant busyness and overwhelming amount of work, but also because a critique is often expressed by critical voices toward the use of contemplative practices in education, whether in schools or in academic research, as being too inward, too private, and simply a personal journey not connected to others. Ruben Habito (1997), in an essay about the connections between meditation and ecology, challenged this notion with the argument that contemplative practices are important to the "deepening of one's mindfulness" (p. 168), and that through practicing mindfulness, a person can gain a stronger presence in the world, what he called "a greater sense of wholeness and at-homeness in ordinary life" (p. 168).

Through efforts to be contemplative, to meditate—if even for a moment—we become present at those places where life is integrated and connected, at those places where knowledge and wisdom are born. In this way, meditation becomes a source of focus, energy, and awareness. We stop being distracted. We can be here, in schools, with children, in life. Habito wrote, "With such an understanding of living the present moment, one lives life and makes decisions in the present in a way that is open to the future and is thereby responsible for it" (p. 171). This is not the fantastical future of the vision of global economic corporations (and Western governments), but the one connected to living and decisions about life today. It is the future emerging through present living relationships. And so it is in this sense that contemplative practices are by no means only inward and personal, but also outward and transformational, rippling out into the world with immeasurable and usually unknowable consequences. We inhabit our lives in time and place, together with others.

Contemplative practices hold us in this present, in this interconnected moment, and deeply in its connection through time. In this purposeful action of slowing down, we orient our hearts toward life, both now and in the future. And through this action, the future might open up and out, unfolding from *here*. A deep engagement and presentness to life—to others, to ourselves, to historical circumstances—materialize future possibilities. In this way, one person's contemplative practice and living are vital to the formation of relationship and community. We sense connections and intuit the ways that boundaries between us are but feeble constructions. Catherine Keller (1986) wrote that:

> we can feel the future forming in ourselves now, for this my present self will be endlessly taken up and reiterated. The future will—if only to the most trivial degree—feel this present. My soul, my body, my world: ongoing, they will have to take me in. So if I learn to feel the subtle movement from past to present, I may begin to discern the transformation of vast relational patterns, personal and social, as they roll through my present. (pp. 246–247).

September 22, 2005

On the east side of our school yard is a little rise, with a few straggly struggling trees, grass, litter. It is not very nice, but we go there anyway. A warm breeze softly blows against our skin. Each child has chosen a comfortable space. Writing and drawing. In their notebooks are pictures and words of trees, of insects, of leaves, of clouds. Everything that isn't on the inside of the school. A child runs to me. She has drawn a seagull flying. Beautiful. Captured in motion. The motion of her smile. What she has seen. The curriculum tells us to learn about plants this year. Outside we are experiencing life, alive with our bodies. The soft roughness of the grass under stomachs. The rustle of the wind in the drying leaves. It is noisier than inside. Inside are the sounds of bells, feet in the halls, the loud heater fan in the corner of our room, the voices of 25 children in a small space. Outside there is room enough for all of us. For the trees, for the birds, for our bodies and minds. I see a child stroking a tree, feeling the bark's ridges under his hands. He finds a crack and worries that the tree will die. No one asks if they are "doing it right" or complains about someone else bothering them in their work. Their bodies are grounded. A child who is never settled inside the school, writes and writes as he lays sprawled widely on the soft grass.

In sensing the ways that curriculum and teaching are bound by daily time, space, circumstance, and indeed part of the "future forming in ourselves now," my work with children in schools takes on a sudden and urgent sense of agency and responsibility. The real feeling of the present moment is the material reality of the children's lives, of their families' lives, of all our lives sharing this earth home. Where does courage come from, to open teaching practice to this present earthly moment and not to an unknown future, or to the global economy, or to children being workers? What we do in our classrooms matters deeply to much more than corporations or governments or stakeholders. It matters to life. A terrifyingly huge thought. Time becomes ecological, glimpsed in the ways that every moment is profoundly connected through all life through all time. Nothing is insignificant. These relations obligate us. Robert Thurman (2004) wrote, "We need to face the probability of the infinite consequentiality of every single thought, word, and deed" (p. 161).

September 9, 2005

It is the seventh day of school. The children are meditating as they paint and draw flowers all day in the Name of Science. They have as much time as they need. Materials are spread around the classroom. Brushes, paints, drawing pencils, pens, paper. We talk about deep looking, about practicing again and again, about going slowly, taking time. They are excited by the aphids on the stems, the aphids crawling across their papers. It is now quiet. And now noisy with chatter and laughter. Now giggling. Now surprised voices praising the work of the next person at the table. Now quiet again. Now at the end of the day the tables and floor are scattered with flowers. The color splash of drying paintings.

Teaching as contemplative action is deliberate. The path that I walk together with children emerges from the place of choosing what to think about, what to focus on, what direction the day-to-day, moment-to-moment work will take. What will occupy children's minds in this space? Robert Thurman (2004) reminded us that everything we do in our lives is already a form of meditation, and therefore that it matters what we choose to think about. I remind myself consciously that there are parts of teaching that I should not allow to occupy my mind and heart. Not thinking too much about preparing for the future. Not thinking too much about high-stakes exams. Not thinking too much about competition, either between children, colleagues, or schools. I can choose what to think about.

September 30, 2005

I meditate today on the ways that schooling practices have tended to separate and distance us from one another. Each child is labeled a "learner." I am labeled a "teacher." We are caught in the Western dualist nightmare where I am supposed to do something to them, rather than something with or for them. Something together.

We are living in a time of intense strain in the global community, not just political and economic violence, but on local and global ecosystems. In the West, we have learned well the habit of disassociating the human from nature, of seeing ourselves as separate and apart. This deep forgetfulness.

The walls of the schools, so often without windows, distance us from the world. We might forget the changing leaves outside, or the weather, or that the world sustains our life and breath. In the schoolyard there are ladybugs that delight the children. They laugh and shout, running back and forth with pleasure and excitement. I vowed to go outside more often this year. So far, we have spent too much time inside our classroom. Too much time engaged in individual work. Not enough time talking together. Not enough time outside. And I wonder, once again, if it is even possible, through contemplative effort and action, to nurture a different intimacy with life and knowledge in this place called school.

To bring mindfulness to the moment of teaching reminds me that it is when I forget the interconnected relations between us, when I see the child as a "skin-encapsulated ego" (Keller 2002, p. 274), as a "separate self," distanced from me, that I am overwhelmed by feelings of anger, incompetence, and helplessness in this work. When I get a momentary glimpse of the fragile connections binding us together in this earthly life, I am suddenly moved. I hear the beauty, thought, depth, brilliance of a child's thoughts. Opening my own heart to them, I am being transformed and moved by this friendship with young people.

To bring mindfulness to the moment of teaching means bringing awareness to what I think and feel. To my fears. It means working hard to give up the desire for absolute control over this classroom space in order to create more spaciousness, more openness, more listening. More room for life. More time for the delicate balances, intimacies, and complexities of this space to emerge.

October 4, 2005

All morning I have been rushing and rushing. There was another organizational staff meeting at lunch, and I have a meeting after school with a parent.

The teachers pass one another in the hall at top speed. We laugh about it in these public places, but in private we talk about how stressed and overwhelmed we feel, how we might burst into tears at any moment.

My hurried trip down the hall toward my classroom to prepare for the afternoon is interrupted by hysterical sobbing at the playground doors. A tiny child is standing there with his eyes squeezed tightly shut. Blood is smeared across his cheek and his lip is puffed out. Between sobs he tells me how he tripped and fell on the playground. I tell him it doesn't look too bad, as I wince at the little bits of sand embedded in his lip and scraped chin. We hold hands and walk slowly down the hall together. We find a quiet place to sit and I clean his face for him. He stops crying. We sit there together until the bell rings for the afternoon to start.

And I realize that I am ready enough for teaching this afternoon.

To bring mindfulness to the moment of teaching is to be able to respond to what is really going on, to life as it presents itself, with all of its surprises from moment to moment. It is preparing for that, not for the supposed future; rather, it brings the future into this moment already here. We do not know what is coming, or if we will be together in the future, but today we are here, together, now, in this place. In this time. We hold lightly to this moment, taking it seriously, knowing that what we do now creates the future. We delight in ladybugs crawling over our arms in the fall sunshine.

October 12, 2005

We are gathered on the carpet reading Jon Muth's (2005) beautiful picture book Zen Shorts *for the sixth time. Slowly. Only a few pages this time. Listening to the poetry in the words. Noticing the details in the illustrations. Talking about the meaning of the stories. Zen stories. Messages meant for hearts. Then the children are finding comfortable spaces around the room. Writing. Deep in thought. We share our words. They merge together, creating a new story between us. Many weeks pass between the first reading and the sixth reading. New insights continue to be formed. A child says, "It took Jon Muth a long time to write this book, so we should take a long time reading it." And so we stay here in this story, an intimate dwelling place we now share.*

Contemplation is a way of gathering insight and stillness together. It brings an urgency into our ordinary, every moment lives—not into extraplanetary life, or wishful thinking, or dreaming of the future. How important this seems in our future-obsessed profession, projecting its fears and anxieties about the lives of children, with extreme pressures on teachers to make children perform to some specific standard at some precise future moment. We are easily distracted by these demands. We might even have an out-of-body experience in these buildings. We lose our grounding in life, time, the earth. We forget that children come as they are.

October 15, 2005

Some days, everything seems to go wrong at school. How is it that I am so easily distracted by the tensions and stresses of this work? I have such lofty ideas, but am often overwhelmed by the external demands. I notice how quickly I slip into that place of "I'm not working hard enough," or "The children aren't learning what they are supposed to," or worse, "The children are not who they are supposed to be."

There are days when I forget that life is a mystery.

Then today, one of them whispers to me in a sad, shy voice, "I'm so hungry." I feel suddenly jolted back to the present, body-slammed away from the place of fearing the government math achievement exam at the end of the year. How did I get there?

Breathe.

It's hard to stay here, in this moment. Hard to bear their hunger, or their hurting bodies, or their grief at losing their homes in wars, in earthquakes, in economic troubles. Perhaps it is easier to fantasize about a pleasant future if we all just work hard enough.

Distractions and stresses arrive suddenly as waves of busyness. It is hard to find time to think. It is hard to find time to be present to children. Tears of frustration burn in my eyes on the way to work and I blink them back. I listen to the world news on the car radio and I think that I do not know how to be a teacher in this world.

Remember that the urgency of life is already here in the very difficult work of moving through a single day, of negotiating relationships and making decisions about what to do next in this teaching work. That is already difficult enough, urgent enough. I struggle to live in the midst of the tensions between the excessive performance demands of this profession and the ways the heart witnesses the truth about the world in the stories of children's lives. In the stories of our own lives. In the story of the earth. I am learning that bringing mindfulness to this moment of teaching means practicing facing this anxiety, loosening my grip, and convincing myself that there is not so much to do after all.

October 10, 2005

It is early Monday morning. The sun is rising, again, as it does—as it has—an enormous pink orange prairie sky. Contemplate the geological time of the earth. The time of the universe. Spectacular overwhelming incomprehensible cosmic time. And the time of the singular individual life. The life of one child in this classroom. In my mind, I try to hold these together as I prepare for teaching today.

To not lose perspective. To remember in those moments when life seems small or trivial, that life is miraculous. Amazing. This interconnected earth. This ecologically connected and arising life where the circulating planet breath moves between all life, from species to species, shared without boundary.

It takes so much effort to hold myself in this place. To try to find the meaning of this work as a teacher within the context of life on this planet, to see our lives in this classroom as part of the arching time of life, and the body space of life. Time spirals and I feel dizzy.

But then I arrive at school, and I see children playing on the playground, their bodies are swinging and they are laughing together. I breathe and orient myself to this day. Just one sun rising. Just one day. Just today. This moment of laughter and love echoing from the climbers. Children from all over the world, speaking dozens of languages, are playing together in peace.

Through contemplation I try to invoke a historical situatedness in these relationships with children. Attempting to understand the complex ways that we came to be here together in this moment matters ethically for how our days emerge in this classroom space we share. Not allowing ourselves to be situated in a future time of wishful thinking and fantasies, I see that our time is already more than enough time. The time of our living presently might already overwhelm us. Some of the children I know have suffered a lot, more than any one human being should in a lifetime. Some of them are exhausted and are in desperate need of a healing place, quiet, joyful, community. To hold them in my heart has meant meditating on suffering. But to linger in those places of pain and sorrow, to try to grasp what they mean for human living and for children's learning lives in schools does not mean dwelling in some kind of hopeless melancholy, but rather finding a place where we can work and be present together in the lives and bodies we truly inhabit. Perhaps it means finding a place where "the task of life becomes to meditatively work toward the happiness of all beings" (Fisher 2002, p. 113).

I wonder what might happen if this were the vision and goal of our schools.

October 10, 2005

The poet David Whyte (1998) warns about making our children too small for the world. How do I not let life become too narrow, too small, to hold their questions, dreams, and experiences? How do I hold all the complexity of the world present in my heart as I open the classroom door in the morning? Like so many classrooms in urban North America, our classroom is an intersection in time and place, a meeting of historical circumstance where human vulnerability and suffering might be revealed and exposed. Children in our class talk about the wars their families have lived through. I tell them quietly that this is a good way to use the word sorrow, which we have recently learned from a story we read together.

I consider the ways that I find myself, here, face to face in this moment with global imperialism, violence, competition over resources, growing ecological instability, and the genocides of the late 20th century. I meditate on all this in an attempt to remember my own implicatedness in these interconnected events, to not become distracted by the pressure to shove math facts into the children's minds. I am trying to understand what is going on. Trying to get the feeling of this present moment. Trying to find stillness with the children. To be with the life and with the math. And for even the math to become meditative, beautiful, and shared between us in wondrous ways.

Perhaps a miracle will grow out of this present situation. Out of all this struggle, suffering, diversity, creativity. Bringing a contemplative attitude to preparing for the day, choosing what to think about, perhaps creates classrooms where life can be lived in many possible and diverse ways, through our relations and the spiraling-out stories of each person here. The ways we influence one another. The spiraling-out story of the earth, the other planets. The spiraling-out history of the universe, of infinite time magnificent and incomprehensible. Perhaps then, the talk about preparing children to succeed in the global economy will fade to a background murmur, not able to enter this place made sacred through our relations.

October 19, 2005

We spend time at our staff meeting talking about how to get our school's test results up. That dizzy feeling is starting again. I feel disoriented. What are we doing here? Who are we competing against? Surely not the tens of millions of children in the world not going to school at all, and not those working literally to the bone to find even meager food for their families, and not the ones orphaned or dying of AIDS in Africa. Surely not them. What are we racing toward and what is the prize at arriving? What are the costs of losing? What happened to working toward the happiness of all beings?

I want to know what it means to learn and teach, and to be together, in a world where there is a Rwandan genocide, a Chernobyl, a September 11th and its violent arching responses. I don't understand what it means to teach and learn in a world without whales, grizzly bears, wild salmon, or monarch butterflies, and so many species, languages and cultures whose names I did not get to learn. Whose names these children will know only as memory, no longer possibility.

I ponder these questions as I prepare for the children to arrive today. So often teaching is represented (and even attempted) as a smooth series of operations and transactions, of simply imparting knowledge, of children moving from activity to activity, from grade to grade. This facade crumbles to dust when Iraq and Bosnia are literally present in our classrooms, in the faces of living, breathing children. I am paralyzed.

Contemplation is not a method, but rather a practice—an everyday practice that can transform the world. It is difficult work. It is about being prepared to meet life moment by moment as it arrives. Being in this moment is hard. Sometimes we feel disappointed. Sometimes overwhelmed. Sometimes also surprised by joy and beauty. This discipline of greeting each moment as it arrives seems somehow opposite to the ways we are often expected to be in schools. It locates us in a larger vision, in a more expansive and generous place-time. It is making space to remember that we are in the midst of all these many relations, always. It is trying to write a different story about what it means to be in schools in the context of what is happening globally, ecologically, politically, economically.

David Whyte (2001) suggested that we might be educating ourselves out of our relations, developing amnesia, a forgetfulness as a result of the speed of our lives: "We forget that our sanity is dependent on a relationship with longer, more patient cycles extending beyond the urgencies and madness of the office" (p.118). Slowing down and being *here* brings us face to face with the implications of ignoring our connectedness. It reminds us of our deep ethical obligations to life over and before politics or economics and the demands they make on our performance as teachers. Through contemplating, we might realize that we have to take a stand, that we have to stand for something. We find courage to face the day-to-day dailyness of our lives, and the enormous ungraspable flow of life and time outside the walls of schools. The economy, the workplace, the factory, and the school are not our true home. Ecologically, the earth is the only true habitat, supporting our lives and relationships in a radical interconnectedness and arising through time. Perhaps, if only for a moment, contemplative practices in education can bring us to this intimate ground, settle us down. If only for a moment.

September 20, 2005

A moment of wonder today. A child, recently arrived in Canada, exclaims that he didn't know! He didn't know, he says, laughing, about the leaves on the trees turning yellow and falling off. The awe in his voice. He didn't know. He is so surprised. Fall has arrived quickly this year. Not much time for leaf-crunching walks. Or for noticing the beauty and loveliness of this place that is our home. We have been collecting and drawing leaves in our sketchbooks, but I wonder if only one person in our class has really noticed them. This child gave us all a gift today.

The time of life. This fleeting fragility. Contemplating mortality. Making friends with time, with contingency, with the ways everything depends on everything else. This, so that we do not waste children's time. We understand that their time is important. This is the only life they have. Holding the moment of a child's life in our minds, we remember that their life-time is precious, that all life, as part of the whole woven time of life, matters in ways we cannot know. This obligation to the world and life is ecological, very serious, very deep. What is at stake is the human species' relationship(s) with all life on earth. Such an understanding of time is in tension with economic time that has governed schools throughout their industrial history. It is ecological and relational time, grounded in the earth. The time that is very old and very deep, grinding down mountains and making sand. While a child sits frustrated in a hard chair in school watching the clock tick slowly, so slowly.

September 25, 2005

Violence erupts in our classroom through words. A suffering and broken child speaks in a loud cruel voice with hatred and malice, in public, about another child who is sitting nearby. He will not stop when asked. The words are spilling out and out, like poison falling from his lips, spreading around the room. Like bombs. Like death. I am overwhelmed with something like sadness. Like anger. I cannot speak. I am the teacher and I could not stop this terror.

Tomorrow we will try again. Try again toward peace. Day by day.

Contemplative teaching moves with time and in space, to the rhythms of life, to what is happening now. It is willing to doubt. To be uncertain. To not know what to do. It is responsive and responsible. Obligated to this time and not an unknowable future. Chogyam Trungpa (2005) wrote that "the practice of meditation is based, not on how we would like things to be, but on what is" (p. 43). He suggested that the only way to relate to the present moment is through "relating with the emotional situations of daily life in a meditative way, by working with them, being aware of them as they come up. Every situation then becomes a learning process" (p. 49). We don't know what is going to "come up" in our classrooms when we resist planning everything ahead of time, resist imposing our structures and desires and wishes on the situations, making space for children's lives and children's bodies. Different paths not yet perceived might open up through children's questions and wonderings. And then new languages and possibilities might emerge, perhaps taking us beyond the language of measurement and accountability, into places that are fresh, surprising, confusing, difficult, and wonderful.

October 28, 2005

I often forget to take time. Time to linger and be still. I forget to give children time to linger and be still. I forget that it takes time to wait for ideas to arrive. I am guilty of hurrying them along, of hurrying myself. I don't even know where we are going. And still I hurry hurry hurry. I can feel their anxiety and mine also. I think they have caught it from me. Teaching is difficult work. But one thing I have learned: Tomorrow is time to try again.

Life spirals out from here and everywhere. Intermingling relations unknowable and mysterious. To bring mindfulness to the moment, discovering again and each day how teaching might be an act of meditative discernment, of wisdom, of love. To remember our connections to the earth, to all life, through time, into the future and to beginnings long in the past. To not forget the weather, the others, all that sustains us. To take time for poetry, for just being together, for listening to the "silence of the world turning" (Domanski, 2002, p. 245). The future of the earth might depend on this. To not be distracted. To remember that our lives are mortal, fragile, lovely. To remember the lives that children have already lived, so that what we do in our classrooms does not erase their time, their experiences, their knowledge, their languages, their own lived connections to the world. To send them a message of courage, peace, and love. To create a place where children can connect to the world, each one, in his or her own way and time. To imagine our work as sacred, these decisions full of infinite consequences. To remember that we are all connected here. That we are all connected here. That we are all connected here. Today.

16. *An Open Letter after a Tough Class and An Afterword to Readers*

David W. Jardine

Over the course of assembling and writing this book, we were involved in a set of four classes with local teachers investigating ecology, hermeneutics and Buddhism as ways to support and find refuge for beautiful classroom work—basically, the themes of this book itself.

Right in the midst of this work together arose a particular insight that ties these three lineages together: a terribly difficult recognition of the falsity of the idea of progress, of "getting somewhere"—ideas deeply precious to education in its aspirations with the young who are inheriting the world and with something of whose care we have been charged. Part of the ensuing lament was my own. After nearly 30 years of working with teachers and students in Calgary, Alberta, I mused that, basically, "things" are exactly the same as they were when I started. We all then began the hard work of considering what to do with such a statement, how not to simply "give up," but how, maybe, to aspire to something different than "getting somewhere." Maybe it is precisely the hope and anxiety and monstrous exaggerations of "getting somewhere" that block and distract attention and that distort our ability to experience and share the real hope that comes from living our lives with the young and their vibrant, difficult, repeated arrivals.

I've elaborated some of the texts that were used in the class and referred to in this email, but otherwise left it alone.

It's offered, here, simply as a note of thanks to readers.

. . . . . .

As a follow up to our wonderful class last night, a passage cited at the end of Chapter Five of the *Pedagogy Left in Peace* [PLIP] (Jardine 2012) book from H.G. Gadamer:

> One has to ask oneself whether the dynamic law of human life can be conceived adequately in terms of progress, of a continual advance from the unknown into the known, and whether the course of human culture is actually a linear progression from mythology to enlightenment. One should entertain a completely different notion: whether the movement of human existence does not issue in a relentless inner tension between illumination and concealment. Might it not just be a prejudice of modern times that the notion of progress that is in fact constitutive for the spirit of scientific research should be transferable to the whole of human living and human culture? One has to ask whether progress, as it is at home in the special field of scientific research, is at all consonant with the conditions of human existence in general. Is the notion of an ever-mounting and self-perfecting enlightenment finally ambiguous?
> from Gadamer, H.G. (1983). *Reason in the age of science*. Cambridge, MA: MIT Press, pp.104–105)

And a passage cited in the introduction to PLIP:

> Insight is more than the knowledge of this or that situation. It always involves an escape from something that had deceived us and held us captive. What [we] learn through suffering is not this or that particular thing [but] insight into … the experience of human finitude. The truly experienced person is one who has taken this to heart. (Gadamer 1989, p. 357).

We hit a really important spot last night, once that each of us has to work out and work through—that phrase "a relentless inner tension between illumination and concealment," is something most Wisdom traditions have taught, that human life is not gradually overcoming its fleeting lot of struggle and impermanence, but only, here and there, coming to learn to live with this lot and still pursue beautiful things.

Someone said last night, and it echoed over and over again: "If we aren't getting anywhere ('continual advance'), why bother at all?" Because, to use Jackie's lovely image [see Chapter 1], miracles still happen. Such moments may not be ever-accumulating, ever-mounting a steady grade "upwards," more and more, in the world—they may simply come and go, here and there, this school and that, seasonally appearing and disappearing, living, thriving, perishing.

But these experiences that we have endured and gathered can, in each of us, "accumulate" and "grow" and affect, then, how each of us moves and lives in this sometimes-sorry world. To quote [David G.] Smith's "Mission" paper (Smith 2006b, p. 105):

> The scholar oriented by the hermeneutic imagination is … interested in … engaging Life hermeneutically [noticed his capitalization], which means trying to understand ever more profoundly what makes life Life, what makes living a living. This is not playing with words: this is asking for the conditions under which it is possible for us to say that we are alive; that our lives are lively, not deadly; and that living seems worthwhile, not just something to be endured until its putative end.

However, being more experienced in understanding these conditions doesn't culminate in knowing more and better than anyone else (expertise). "Experience has its proper fulfillment not in definitive [amassed] knowledge but in the openness to [new] experience[s] that is made possible by experience itself." (Gadamer 1989, p. 355).

Through the course of becoming experienced in this way of experiencing the world and practicing it ourselves and with our students, we don't end up with an ever-mounting success in ever-more schools. The course of our work is not fulfilled by progressively "more and more" widespread miraculousness, but in an ever increasing, quiet and generous confidence in the truth of miracles *despite* the successes or failures of such matters in the world. This misplaced hope that the world might get increasingly better if we just try hard enough or work long enough is only a path to despair. To quote Smith's "Wisdom Responses" paper (in press):

> Pedagogically, inducting children [our ourselves!] into a belief that life is a matter of will, and will power, under the guise of clichés like "You can do anything in life that you want to do, if you work hard, and put your mind to it"—all this can be a recipe for despair in the face of failed dreams. Similarly, preoccupations with goal-setting, curriculum-by-objectives etc., are not ill-advised in themselves but quickly become so if they evolve into blinkered constraints against the fullness of life's beckoning."

Hermeneutic work, then, is "fulfilled" and "fulfilling" (in us and in our students) by resulting in an increasing readiness, openness and willingness to face the realities of the world—an "openness to new experiences." This openness to new experiences means that I am willing to ask of what comes to meet me whether it is worth while (see "Fideles paper" [Chapter 7]) or whether it is merely another distraction. This is why hermeneutics is not linked to anonymously amassing knowledge (the knowledge pursued in the natural sciences), but is necessarily about "self-formation" (German *Bildung*)—I myself *become someone* as a result of the way I make through the world and through my forming life. This is the big distinction that happened at the end of the 19th century between the natural sciences and the "human sciences" (see Smith's "Mission" paper [2006b] regarding Dilthey and Schliermacher)—in the human sciences, like education, I know that, in many complex ways, what I am studying is *myself* and, in pursuing such a study, I become someone (*Bildung*) different than I had been before—I become "worldly" in my understanding of the world and myself. I become increasingly aware of my life in the life world.

Bildung, generously understood, is not simply an inner repose or becoming "cultured", but a readiness to more honestly and open-heartedly meet the troubles of the world. Our wonderful work with students makes us better able to speak up in the face of those things that might intrude, to name these intrusive forces and false hopes, and to make a case, out in the world, for what we do.

Last night I mentioned a CBE [Calgary Board of Education] upper level person (so to speak) at a meeting—someone with a doctorate, someone published along the lines of the things we've been reading, who ran an exquisite classroom for years, etc.—who fell silent in the face of Ministry talk about interventions in Kindergarten for literacy skills testing. She didn't say a word at that meeting. I guess that all this is partially what was guiding my own thoughts yesterday.

I'm really not interested in being doomy and gloomy about the troubles we are in, but rather, I hope, clear minded and unafraid to face them and work them out, untangle them.

So, check out Chapter 11 of PLIP (Jardine 2012f, pp. 228–9):

> a pedagogy left in peace does not arrive simply by cultivating a still inwardness (Sanskrit: *Samadhi*) separate from the [troubles of] world. Such stillness is a necessary condition of our healing, but it is not sufficient condition. A pedagogy left in peace also invites and requires the cultivation of wisdom (Sanskrit: *Vipassana*) about the world.

There is a lovely passage in there cited Ajhan Chah (2001, p. 91) on page 229, where he contrasts the sort of peace that can arise from stillness and undistractedness with another sort of peace that is essential—a peace that arises from wisdom:

> The peace that comes from wisdom is distinctive, because when the mind withdraws from tranquility [and moves outwards into the surrounding troubles of the world, like that whole first part of Smith's Wisdom Responses paper], the presence of wisdom makes it unafraid. With such energy, you become fearless. Now you know phenomena as they are and are no longer afraid.

This is why I believe that the second half of Smith's "Wisdom" (2004) paper (where he talks about the Wisdom traditions and we can feel an affinity to our own work) makes us better able to "face" the first half (where he talks about the terrible troubles of global forces and neo-conservativism). It makes me more courageous.

So hermeneutics has this double aspect—undistractness and wisdom. Up to a point, our ability to do wonderful things with kids and for each other is fundamental to this work we are describing. Gadamer says outright that human life *is* hermeneutic—open to interpretation, full of interrelatedness and ambiguity and abundance—and, therefore, doing rich inquiry in the classroom is in line with this understanding of being human (again, back to the natural sciences/human sciences debate, back to a debate about the infusion of F. W. Taylor's work into education—where gaining knowledge be came pictured as assembling an object according to rules akin to natural scientific methodologies).

But part of hermeneutics is also doing the tough work of providing a "free space" in which we can then unravel those things that have come up against this understanding of being human; we can unravel and erode those things that distract us. Part of our job is to not turn away from the bad news and simply retreat. This leaves our kids and us unprepared to meet the world and decode what is happening to us.

Sad to say, however, that as we erode these institutional and personal and cultural and economic and political things that distort and demean our work, they rise up all over again: " A relentless inner tension between illumination and concealment." Our stillness allows us to see this happening and not fall for it, allows us to name it in great detail, to bypass it, to speak differently, to speak up (see that McLaren article of Smith's) while not simply exhausting ourselves at this breach in the hope of changing the world forever. But our stillness and refuge in the miraculous things we want for us and students is not enough.

As King David says in the Psalms, human life is lived out in a "land of shadows." As that other David suggests, this is what "makes life Life."

So we've landed on the toughest insight possible here. The work of "inquiry in the classroom" really *is* operating from a deep and difficult insight. All that talk of Market Economy, the efficiency movement, bureaucratization, more forms, more distraction—what is it fundamentally distracting us *from*? "The experience of human finitude. The truly experienced person is one who has taken this to heart" (Gadamer 1989, p. 357). This places an unfixable, un-fillable, unutterable "gap" right in the middle of what we do.

To quote Hannah Arendt (1969, pp. 192–3):

> To preserve the world against the mortality of its creators and inhabitants it must be constantly set right anew. The problem is simply to educate in such way that a setting-right remains actually possible, even though it can, of course, never be assured.

I go, then, back to Jackie's paper [see Chapter 1], that, in light of these insight, it is even *more* miraculous that miraculous things can happen, because it is *never* assured. I might even venture further, that it is only *because* these arrivals of insight and worthwhileness are impermanent, rare and finite that we can truly love them and dedicate ourselves, as best we can, again and again every September, to their arrival.

References

Abram, David (1996). *The spell of the sensuous: Language in a more-than-human world.* New York: Pantheon Books.

Alexievich, S. (2005). *Voices from Chernobyl: The oral history of a nuclear disaster.* (K. Gessen, Trans.). London, UK: Dalkey Archive Press.

Anderson, H. (2006). Teabags. On-line at: http://mothertongued.com/hortensia/teabags.htm. No longer accessible. Last accessed June 2007. See next citation for further information.

Anderson, H. (2007). *The plenitude of emptiness: A collection of the Asian poetic form known as Haibun.* Available on-line at http://hortensiaanderson.blogspot.com.

Arendt, H. (1969). *Between past and present: Eight exercises in political thought.* New York: Penguin Books.

Aristotle (1941). *The basic works of Aristotle.* R. McKeon, Trans. New York: Random House.

Ayres, L. (1909). *Laggards in our schools.* New York—pdf version is available on line at: http://www.archive.org/details/laggardsinoursch00ayrerich.

Ayres, L. (1915). *A measuring scale for ability in spelling.* Available on line at: donpotter.net.

Bakken, D. (2003). Ink that echoes. *The marquee.* Available on-line at: http://www.bisbeemarquee.com/www/2003/1116/c07.php. Accessed August 21, 2007.

Beresford-Kroeger, D. (2003). *Arboretum America: A philosophy of the forest.* Ann Arbor: The University of Michigan Press.

Berk, A. & Long, L. (2012). *Nightsong.* New York: Simon and Schuster Books for Young Readers.

Berman, M. (1983). *The reenchantment of the world.* New York: Bantam Books.

Berry, W. (1983). *Standing by words.* San Francisco: North Point Press.

Berry, W. (1986). *The unsettling of America: Essays in culture and agriculture.* San Francisco: Sierra Club Books.

Berry, W. (2002). People, land and community. In W. Berry (2002). *The art of the common place.* Washington D.C.: Counterpoint, pp. 182–194.

Bethune, R. (2002). To translate is to betray? *ArtTimesJournal.* Available on-line at: http://www.arttimesjournal.com/theater/totranslate.htm. Accessed August 18, 2007.

Bly, R. (1988). *A little book on the human shadow.* New York: Harper and Row.

Bobbitt, F. (1918). *The curriculum*. Boston: Houghton Mifflin.

Bobbitt, F. (1924). *How to make a curriculum*. Boston: Houghton Mifflin.

Boyle, D. (2006). The man who made us all work like this .... *BBC History Magazine*, June 2003. Accessed August 5, 2009 at: http://www.david-boyle.co.uk/history/frederickwinslowtaylor.html.

Braverman, H. (1998). *Labor and monopoly capital: The degradation of work in the twentieth century*. New York: Monthly Review Press.

Butler, J. (2004). Betrayal's felicity. *Diacritics*. *34*(1), Spring 2004, pp. 82–87.

Callahan, R. (1964). *America, education and the cult of efficiency*. Chicago: University of Chicago Press.

Caputo, J. (1982). *Heidegger and Aquinas: An essay on overcoming metaphysics*. New York: Fordham University Press.

Carroll, L. (Originally published 1871). *Through the looking-glass and what Alice found there*. Designed and published by PDFreeBooks.org. Full text accessed on-line August 9th 2012 from www.google.ca/books.

Carruthers, M. (2003). *The craft of thought: Meditation, rhetoric, and the making of images, 400–1200*. Cambridge, UK: Cambridge University Press.

Carruthers, M. (2005). *The book of memory: A study of memory in medieval culture*. Cambridge, UK: Cambridge University Press.

Carruthers, M. & Ziolkowski, J. (2002). *The medieval craft of memory: An anthology of texts and pictures*. Philadelphia: University of Pennsylvania Press.

Carter, C. (2012). Teacher: 'I wanted to be the last thing they heard, not the gunfire.' Accessed on-line from *CNN News On-Line*, U.S. Edition, December 16th, 2012 at: http://www.cnn.com/2012/12/15/us/connecticut-childrens-story/index.html.

Chah, A. (2001). *Being Dharma: The essence of the Buddha's teachings*. Boston MA: Shambhala Press.

Chambers, C. (2012). Spelling and other illiteracies. In C.M. Chambers, E. Hasebe-Ludt, C. Leggo, & A. Sinner (Eds.). *A heart of wisdom: Life writing as empathetic inquiry* (pp. 183–189). New York, NY: Peter Lang.

Chandrakirti (2002). *Introduction to the Middle Way: Chandrakirti's Madhyamakavatara with commentary by Jamgon Mipham*. Boston: Shambhala.

Charters, W. (1923). *Curriculum construction*. New York: Macmillan.

Chodron, P. (1997). *When things fall apart: Heart advice for difficult times*. Boston: Shambhala.

Clifford, J. & Marcus, G. (1986). *Writing culture: The poetics and politics of ethnography*. Berkeley, CA: University of California Press.

Dawson, C. (1998). Translator's introduction to H.G. Gadamer (1998). *Praise of theory: Speeches and essays* (pp. xv–xxxviii). New Haven, CT: Yale University Press.

Derrida, J. (2001). *The work of mourning*. Chicago: University of Chicago Press.

Descartes, R. (1955). Discourse on method. In *Descartes selections* (pp. 249–266). New York: Charles Scribner's Sons.

Domanski, D. (2002). The wisdom of falling. In T. Bowling (Ed.), *Where the words come from: Canadian poets in conversation* (pp. 244–255). Roberts Creek, BC, Canada: Nightwood Editions.

Domanski, D. (2010). *All our wonder unavenged*. London, ON: Brick Books.

Dorrie, D. (2007). *How to cook your life: With Zen chef Edward Espe Brown*. Toronto: Mongrel Media.

Dudjom Rinpoche & Dorje, J.Y. (2011). *A torch lighting the way to freedom: Complete instructions on the preliminary practice of the profound and secret heart essence of the Dakini.* Boston: Shambhala.

Double Birdie (2005). Three of six: A globetrotting American talks current events, pop culture, and theology. Available online at: http://threeofsix.blogspot.com/2005/07/to-translate-is-to-betray.html. Accessed August 8, 2007.

DuFour, R. & Eaker, R. (1998). A new model: The professional learning community. Professional learning communities at work: Best practices for enhancing student achievement. From *A new model: The professional learning community.* The Eisenhower National Clearinghouse for Mathematics and Science Education (ENC). Accessed on line at: http://www.myeport.com/published/t/uc/tucson73/collection/1/4/upload.doc.

Eliade, M. (1968). *Myth and reality.* New York: Harper & Row.

Evernden, N. (2002). Interview with Neil Evernden. In D. Jensen, *Listening to the land: Conversations about nature, culture, and eros* (pp. 112–121). San Francisco: Sierra Club Books.

Ferreiro, E. (2003). *Past and present of the verbs to read and to write: Essays on literacy* (M. Fried, Trans.). Toronto, Ontario, Canada: Groundwood Books.

Fisher, A. (2002). *Radical ecopsychology: Psychology in the service of life.* Albany: State University of New York Press.

Ford, H. (2007). *My life and work.* New York: Cosimo Books.

Friesen, S. (2010). Uncomfortable bedfellows: Discipline-based inquiry and standardized examinations. *Teacher Librarian: The Journal for School Library Professionals.* October 2010. Accessed on-line January 29, 2011 at: http://www.encyclopedia.com/Teacher+Librarian/publications.aspx?pageNumber=1.

Friesen, S. & Jardine, D. (2009). On field(ing) knowledge. In S. Goodchild & B. Sriraman, eds. *Relatively and philosophically E[a]rnest: Festschrifte in honour of Paul Ernest's 65th birthday. The Montana mathematics enthusiast: Monograph series in mathematics education* (pp. 149–175). Charlotte NC: Information Age Publishing.

Friesen, S. & Jardine, D. (2010). 21st century learners and learning. A report prepared for the Western and Northern Canadian Curriculum Protocol for Collaboration in Education/Protocole de l'Ouest et du Nord Canadiens de Collaboration Concernant L'Education. Accessed on-line January 20th, 2010 at: http://education.alberta.ca/media/1087278/wncp%2021st%20cent%20learning%20(2).pdf.

Friesen, S. & Jardine, D. (2011). *Guiding principles for WNCP/PONC curriculum framework projects.* A report prepared for the Western and Northern Canadian Curriculum Protocol for Collaboration in Education/Protocole de l'Ouest et du Nord Canadiens de Collaboration Concernant L'Education. Accessed on-line March 20, 2011 at: http://www.education.gov.sk.ca/Default.aspx?DN=0583e277–654d-45c5–98fec791c85a118b.

Friesen, S., Jardine, D. & Gladstone, B. (2010). The first thunderclap of spring: An invitation into Aboriginal ways of knowing and the creative possibilities of digital technologies. In C. Craig & L. F. Deretchin eds. *Teacher education yearbook XVIII: Cultivating curious and creative minds: The role of teachers and teacher educators.* Lanham, MD: Scarecrow Education, pp. 179–199.

Gadamer, H.G. (1970). Concerning empty and ful-filled time. *Southern Journal of Philosophy,* Winter 1970, pp. 341–353.

Gadamer, H.G. (1977). *Philosophical hermeneutics.* Berkeley, CA: University of California Press.

Gadamer, H.G. (1983). *Reason in the age of science.* Cambridge, MA: MIT Press.

Gadamer, H.G. (1986). The idea of the university—Yesterday, today, tomorrow. In Dieter Misgeld & Grahame Nicholson, eds. and trans. *Hans-Georg Gadamer on education, poetry, and history: Applied hermeneutics* (pp. 47–62). Albany, NY: SUNY Press.

Gadamer, H.G. (1989). *Truth and method.* (J. Weinsheimer, trans.). New York: Continuum Books.

Gadamer, H.G. (1994). *Heidegger's ways.* Boston: MIT Press. New Press.

Gadamer, H.G. (2001). *Gadamer in conversation: Reflections and commentary.* R. Palmer, ed. and trans. New Haven, CT: Yale University Press.

Gadamer, H.G. (2007). From word to concept: The task of hermeneutic philosophy. In R.E. Palmer, ed. *The Gadamer reader: A bouquet of later writings* (pp. 108–120). Evanston IL: Northwestern University Press.

Gadamer, H.G. (2007a). Aesthetics and hermeneutics. In R.E. Palmer, ed. *The Gadamer reader: A bouquet of the later writings* (pp. 124–131). Evanston, IL: Northwestern University Press.

Gadamer, H.G. (2007b). The artwork in word and image: "So true, so full of Being!" In R.E. Palmer, ed. *The Gadamer reader: A bouquet of the later writings* (pp. 192–226). Evanston, IL: Northwestern University Press.

Gadamer, H.G. (2007c). Hermeneutics as practical philosophy. In R.E. Palmer, ed. *The Gadamer reader: A bouquet of the later writings* (pp. 227–246). Evanston, IL: Northwestern University Press.

Gadamer, H.G. (2007d). Hermeneutics as a theoretical and practical task. In R.E. Palmer, ed. *The Gadamer reader: A bouquet of the later writings* (pp. 246–265). Evanston, IL: Northwestern University Press.

Gadamer, H.G. with Dutt, C. (2001). Aesthetics. In R.E. Palmer, ed., *Gadamer in conversation: Reflections and commentary* (pp. 61–77). New Haven, CT: Yale University Press.

Gatto, J. (2006). The national press attack on academic schooling. Available on-line at: http://www.rit.edu/~cma8660/mirror/www.johntaylorgatto.com/chapters/9d.htm.

Gilham, C. (2012). The privileges chart in a behavior class: Seeing the power and complexity of dominant traditions and unconcealing trust as basic to pedagogical relationships. *Journal of Applied Hermeneutics.* Accessed on-line December 15th, 2012 at: http://jah.synergiesprairies.ca/jah/index.php/jah/article/view/15.

Gilham, C. (2012a). From the "science of disease" to the "understanding of those who suffer": The cultivation of an interpretive understanding of "behavior problems" in children. *Journal of Applied Hermeneutics.* Accessed on-line December 15th, 2012 at: http://jah.synergiesprairies.ca/jah/index.php/jah/article/view/33.

Gilmore, M. (2012). Bob Dylan on his dark new album, 'Tempest': Dylan breaks down his apocalyptic (and sometimes sweet) 35th studio LP. Accessed on-line August 7, 2012 at: http://www.rollingstone.com/music/news/bob-dylan-on-his-dark-new album-tempest-20120801.

Goyette, S. (1998). *The true names of birds.* London, ON: Brick Books.

Gray, J. (2001). *False dawn: The delusions of global capitalism.* New York: The New Press.

Greene, M. (2001). *Variations on a blue guitar: The Lincoln Center Institute lectures on aesthetic education.* New York, NY: Teachers College Press.

Grimm, J. & Vaast, D. (2011). Glendale as been Karshed. *One World in Dialogue. 1*(1), pp. 37–47.
Grondin, J. (2003). *Hans-Georg Gadamer: A biography.* New Haven, CT: Yale University Press.
Gurria-Quintana, A. (2006). *FT.com [Financial Times], Arts and Weekend, Books.* Available on-line at http://www.ft.com/cms/s/a6787988–5f43–11db-a011 0000779e2340. html. Published October 20 2006 16:47. Accessed June 2007.
Habito, R. (1997). Mountains and rivers and the great earth: Zen and ecology. In M. E. Tucker & D. R. Williams (eds.). *Buddhism and ecology: The interconnection of Dharma and deeds* (pp. 187–218). Cambridge, MA: Harvard University Press.
Hahn, T.N. (1986). *The sun my heart.* Berkeley, Parallax Press.
Hahn, T.N. (2008). *The world we have: A Buddhist approach to peace and ecology.* Berkeley, California: Parallax Press.
Hahn, T.N. (2012). The heart Sutra: *Prajnaparamita Hrdaya Sutra.* In Thich Nhat Hahn (2012). *Awakening of the heart: Essential Buddhist Sutras and commentaries.* Berkeley CA: Parallax Press, pp. 407–442.
Hamill, S. & Kaji, A. (2000). *The sound of water: Haiku by Basho, Buson, Issa, & other poets.* Boston: Shambhala.
Hamilton, V., & Moser, B. (illus.) (1988). *In the beginning: Creation stories from around the world.* New York: Harcourt Brace Jovanovich.
Hardon, J. (1985). *Pocket Catholic dictionary.* New York: Doubleday.
Heidegger, M (1962). *Being and time.* New York: Harper and Row.
Heidegger, M. (1968). *What is called thinking?* New York: Harper and Row.
Heidegger, M. (1971). A dialogue on language between a Japanese and an inquirer. In M. Heidegger (1971). *On the way to language* (pp. 1–12). New York: Harper and Row.
Heidegger, M. (1977a). The origin of the work of art. In *Basic writings* (pp. 143–188) New York: Harper and Row Publishers.
Heidegger, M. (1977b). Letter on humanism. In *Basic writings.* (pp. 189–242). New York: Harper and Row Publishers.
Heidegger, M. (1978). *The metaphysical foundations of logic.* Bloomington: Indiana University Press.
Hillman, J. (1983). *Healing fiction.* Barrytown, NY: Station Hill Press.
Hillman, J. (1998). *Inter views: Conversations with Laura Pozzo on psychotherapy, biography, love, soul, the gods, animals, dreams, imagination, work, cities, and the state of the culture.* Dallas: Spring Publications.
Hillman, J. (2004). *A terrible love of war.* New York, NY: The Penguin Press.
Hillman, J. (2005). Notes on opportunism. In James Hillman (2005). *Senex and Puer* (pp. 96–112). Putnam, CT: Spring Publications.
Hillman, J. (2006). Anima Mundi: Returning the soul to the world. In James Hillman (2006). *City and soul* (pp. 27–49). Putnam, CT: Spring Publications.
Hillman, J. (2006a). Loving the world anyway. In James Hillman (2006). *City and soul* (pp. 128–130). Putnam, CT: Spring Publications, Inc..
hooks, b. (2000). *All about love: New visions.* New York: HarperCollins.
Hove, P. (1996). The face of wonder. *Journal of Curriculum Studies 28*(4), pp. 437–462.
Huizinga, J. (1955). *Homo ludens: A study of the play element in culture.* Boston: The Beacon Press.
Husserl, E. (1969). *Ideas toward a pure phenomenology.* New York: Humanities Press.

Husserl, E. (1970). *The crisis of European science and transcendental phenomenology.* Evanston, IL: Northwestern University Press.

Husserl, E. (1970a). *Cartesian meditations.* The Hague: Martinus Nijhoff.

Illich, I. (1992). *In the mirror of the past: Lectures and addresses 1978–1990.* New York: Marion Boyars.

Illich, I. (1993). *In the vineyard of the text: A commentary on Hugh's Didascalicon.* Chicago: University of Chicago Press.

Illich, I. (1998). The cultivation of conspiracy. Accessed November 1, 2012 at: www.davidtinapple.com/illich.

Illich, I. & Sanders, B. (1988). *ABC: The alphabetization of the popular mind.* Berkeley, CA: North Point Press.

Ingram, C. (1990). *In the footsteps of Gandhi: Conversations with spiritual social activists.* Berkeley, CA: Parallax Press.

Jardine, D. (2000). "Even there the gods are present". In Jardine, D. (2000). "*Under the tough old stars": Ecopedagogical essays* (pp. 205–214). Brandon, VT: Psychology Press / Holistic Education Press.

Jardine, D. (2008). "The stubborn particulars of grace." In D. Jardine, S. Friesen & P. Clifford (2008). *Back to the basics of teaching and learning: Thinking the world together* (pp. 143–152), 2nd edition. New York: Routledge.

Jardine, D. (2012). *Pedagogy left in peace: On the cultivation of free spaces in teaching and learning.* New York: Continuum Books.

Jardine, D. (2012a). Sickness is now "out there." In D. Jardine (2012), *Pedagogy left in peace* (pp. 73–90). New York: Continuum Books.

Jardine, D. (2012b). "A hitherto concealed experience that transcends thinking from the position of subjectivity." In D. Jardine (2012). *Pedagogy left in peace: On the cultivation of free spaces in teaching and learning* (pp. 91–112). New York: Continuum Books.

Jardine, D. (2012c). Figures in Hell. In Jardine, D. (2012). *Pedagogy left in peace: On the cultivation of free spaces in teaching and learning* (pp. 133–144). New York: Continuum Books.

Jardine, D. (2012d). "Youth need images for their imaginations and for the formation of their memory." In D. Jardine (2012). *Pedagogy left in peace: On the cultivation of free spaces in teaching and learning* (pp. 159–172). New York: Continuum Books.

Jardine, D. (2012e). On the while of things. In D. Jardine. *Pedagogy left in peace: On the cultivation of free spaces in teaching and learning* (pp. 173–192). New York: Continuum Books.

Jardine, D. (2012f). "Take the feeling of letting go as your refuge." In D. Jardine. *Pedagogy left in peace: On the cultivation of free spaces in teaching and learning* (pp. 217–230). New York: Continuum Books.

Jardine, D., Clifford, P. & Friesen, S. (2006). *Curriculum in abundance.* Mahwah, NJ: Lawrence Erlbaum and Associates.

Jardine, D., Clifford, P., & Friesen S. (2008). *Back to the basics of teaching and learning: Thinking the world together*, 2nd edition. New York: Routledge.

Jardine, D. & Naqvi, R. (2012). Learning not to speak in tongues. In D. Jardine (2012). *Pedagogy left in peace* (pp. 193–216). New York: Continuum Books.

Kanigel, R. (2005). *The one best way: Frederick Winslow Taylor and the enigma of efficiency.* Cambridge, MA: The MIT Press.

Keller, C. (1986). *From a broken web: Separation, sexism, and self.* Boston: Beacon Press.

Keller, C. (1996). *Apocalypse now and then: A feminist guide to the end of the world.* Boston: Beacon Press.

Keller, C. (2000). The lost fragrance: Protestantism and the nature of what matters. In H. Coward & D.C. Maguire (eds.), *Visions of a new earth: Religious perspectives on population, consumption and ecology* (pp. 73–93). Albany, NY: State University of New York Press.

Keller, C. (2002). Catherine Keller. In D. Jensen (Ed.), *Listening to the land: Conversations about nature, culture, and eros* (pp. 273–281). San Francisco: Sierra Club Books.

Keller, C. (2003). *Face of the deep: A theology of becoming.* London: Routledge.

Kostash, M. (2003). Reading the river: The North Saskatchewan in myth, prose and poetry. *Canadian Geographic 123*(6), 50–62.

Lawson, M. (2003). *Crow Lake.* Toronto, Vintage Canada (Random House of Canada Limited).

Leonard, H. (2010). *Jazz.* London: Bloomsbury Press.

Liang-Chieh (1986). *The record of Tung-Shan.* Honolulu: University of Hawai'i Press.

Lobsang Gyatso (2006). Commentary to Tsong-kha-pa (2006). *The harmony of emptiness and dependent-arising.* New Delhi: Library of Tibetan Works and Archives.

Loy, D. (1993). Indra's postmodern net. *Philosophy East and West.* Vol. *43*, #3, July 1993, pp. 481–510.

Macy, J. (2000). *Widening circles: A memoir.* Gabriola Island, BC: New Society Publishers.

McGinnis, S. (2008). Province urged to scrap high-stakes tests. *Calgary Herald*, September 8, 2008, p. B5.

Morrison, B. (2000). *The justification of Johann Gutenberg.* Toronto: Random House of Canada.

Muth, J. (2005). *Zen shorts.* New York: Scholastic Press.

Nietzsche, F. (1975). *The will to power.* New York: Random House

Nishitani, K. (1982). *Religion and nothingness.* Berkeley, CA: University of California Press.

On-Line Etymological Dictionary (OED). Accessed June 9th, 2012 at: www.etymology-on-line.com.

Palmer, P. (1993). *To know as we are known: Education as a spiritual journey.* 2nd ed. San Francisco: Harper.

Paragamian, A. (Director). (2000). *2000 and none.* [Film]. Pandora Film.

Parry, W. (2010). Age confirmed for 'Eve,' mother of all humans. *Live Science.* Retrieved from: http://www.livescience.com.

Patrul Rinpoche (1998). *The words of my perfect teacher.* Boston: Shambhala Press.

Pelden, K (2010). *The nectar of Manjushri's speech: A detailed commentary on Shantideva's* Way of the Boddhisattva. Boston: Shambhala.

Peterson, Roger Tory (1980). *A field guide to the birds east of the Rockies.* 4th edition. Boston: Houghton Mifflin Company.

Plato (2012a). Phaedrus. In *A Plato reader: Eight essential dialogues* (pp. 209–266). C.D.C. Reeve, ed. Cambridge MA: Hackett Publishing Co.

Plato (2012b). Symposium. In *A Plato reader: Eight essential dialogues* (pp. 153–208). C.D.C. Reeve, ed. Cambridge MA: Hackett Publishing Co.

Ree, J. (2000). *I see a voice.* London: Flamingo.

Reibstein, M. (2008). *Wabi Sabi.* Young, E. (Illus.). New York: Little, Brown and Company.

Reston, J. (2005). *Dogs of God: Columbus, the Inquisition and the defeat of the Moors.* New York: Doubleday.

Ricoeur, P. (1970). *Freud and philosophy.* New Haven, CT: Yale University Press.

Rodé, M. A. (2012). How to grow a mandala. *Undivided: The Online Journal of Nonduality and Psychology, 1*(3). Retrieved from: http://undividedjournal.com/2012/12/06/how-to-grow-a-mandala/.

Rodgers, G. (1983). To urge the repetition of the Ayres' spelling tests of 1914–15 to confirm the existence of massive present-day reading disability and to establish its cause and cure. Available on line at: www.donpotter.net.

Rodgers, G. (1984). Historical introduction to Leonard P. Ayres' *A measuring scale for ability in spelling (1915).* Prepared by Donald L. Potter, July 21, 2004, from materials written by Geraldine Rodgers on December 30, 1984. Available on line at www.donpotter.net.

Rosenberg, R. & Hancock, M. (2001). *Exclusions and awakenings: The life of Maxine Greene.* New York, NY: Hancock Productions.

Ross, D. (1999). *The gift of kinds [the good in abundance]: An ethic of earth.* Albany, NY: SUNY Press.

Ross, S. (2004). Gadamer's late thinking on *Verweilen. Minerva—An Internet Journal of Philosophy,* Vol. *8.* Access on-line July 19, 2010 at: http://www.ul.ie/~philos/vol8/gadamer.html.

Ross, S. (2006). The temporality of tarrying in Gadamer. *Theory, Culture & Society.* Vol. 23(1): 101–123.

Ross, S. & Jardine, D. (2009). Won by a certain labour: A conversation on the while of things. *Journal of the American Association for the Advancement of Curriculum Studies.* Accessed July 21, 2009 at: http://www.uwstout.edu/soe/jaaacs/Vol5/Ross_Jardine.htm.

Ruether, R.R. (2002). Ecofeminism: Symbolic and social connections of the oppression of women and the domination of nature. In C. Adams (ed.). *Ecofeminism and the sacred* (pp. 13–23). New York, NY: The Continuum Publishing Company.

Sato, H. (1983). *One hundred frogs: From Renga to Haiku to English.* New York: Weatherhill.

Schopenhauer, A. (1963). *The world as will and representation.* New York: Dover Books.

Sendak M. (1988). *Where the wild things are.* New York: HarperCollins.

Shantideva (2006). *The way of the Bodhisattva.* Boston: Shambhala.

Shepard, P. (2002). Interview with Paul Shepard. In D. Jensen, *Listening to the land: Conversations about nature, culture and eros* (pp. 248–259). San Francisco: Sierra Club Books.

Shirane, H. (1996). *Traces of dreams: Landscape, cultural Memory and the poetry of Basho.* Stanford, CA: Stanford University Press.

Shiva, V. (1997). *Biopiracy: The plunder of nature and knowledge.* Boston: South End Press.

Smith, D. G. (1999). *Pedagon: Interdisciplinary essays on pedagogy and culture.* New York: Peter Lang Publishing.

Smith, D. G. (2006). *Trying to teach in a season of great untruth: Globalization, empire and the crises of pedagogy.* Rotterdam: Sense Publishing.

Smith, D. G. (2006a). Introduction. In D.G. Smith, *Trying to teach in a season of great untruth: Globalization, empire and the crises of pedagogy* (pp. xxi–xxvii). Rotterdam: Sense Publishing.

Smith, D. G. (2006b). The mission of the hermeneutic scholar. In D.G. Smith, *Trying to teach in a season of great untruth: Globalization, empire and the crises of pedagogy* (pp. 105–115). Rotterdam: Sense Publishing.

Smith, D. G. 2014. Wisdom responses to globalization: The pedagogic context. In W. Pinar (Ed.)., *The international handbook of curriculum research* (pp. 45–59) (2nd ed.). New York: Routledge.

Snyder, G. (1980). *The real work.* San Francisco: New Directions Books.

Snyder, G. (1990). The place, the region and the commons. In *The practice of the wild* (pp. 27–51). Berkeley, CA: Counterpoint Books.

Snyder, G. (1990a). Survival and sacrament. In *The practice of the wild* (pp. 175–185). Berkeley, CA: Counterpoint Books.

Sopa, L. (2004). *Steps on the path to enlightenment.* Vol. 1. Somerville, MA: Wisdom Publications.

Sopa, L. (2005). *Steps on the path to enlightenment.* Vol. 2. Somerville, MA: Wisdom Publications.

Sopa, L. (2008). *Steps on the path to enlightenment.* Vol. 3. Somerville MA: Wisdom Publications.

Stephenson, N. (2008). *Anathem.* New York: William Morrow.

Stock, B. (1983). *The implications of literacy: Written language and models of interpretation in the eleventh and twelfth centuries.* Princeton, NJ: Princeton University Press.

Taylor, F. W. (1903) *Shop Management [Excerpts].* Accessed on line August 14, 2010 at: http://www.marxists.org/reference/subject/economics/taylor/shop-management/abstract.htm.

Taylor, F. W. (1911). *Scientific management, comprising shop management, the principles of scientific management and testimony before the special house committee.* New York: Harper & Row.

Thomas, D. (1954a). Reminiscences of childhood. In D. Thomas (1954). *Quite early one morning.* New York: New Directions Paperbook, pp. 3–8.

Thomas, D. (1954b). A child's Christmas in Wales. In D. Thomas (1954). *Quite early one morning.* New York: New Directions Paperbook, pp. 14–21.

Thompson, W. I. (1981). *The time falling bodies take to light: Mythology, sexuality and the origin of culture.* New York: St. Martin's Press.

Thurman, R. (2004). *Infinite life: Seven virtues for living well.* New York: Riverhead Books.

Thursby, K. (2010). Leonard's smoke-filled images of jazz greats in dark clubs documented a musical era. *Los Angeles Times*, August 16, 2010. Accessed on-line September 6, 2011 at: http://articles.latimes.com/2010/aug/16/local/la-me-herman-leonard-20100816.

Trungpa, C. (1991). *Meditation in action.* Boston: Shambhala.

Trungpa, C. (2005). *The sanity we are born with: A Buddhist approach to psychology.* (C. R. Gimian, Ed. & Comp.). Boston: Shambhala.

Tsong-kha-pa (2000). *The great treatise on the stages of the path to enlightenment.* Volume One. Ithaca, NY: Snow Lion Publications.

Tsong-kha-pa (2002). *The great treatise on the stages of the path to enlightenment.* Volume Three. Ithaca, NY: Snow Lion Publications.

Tsong-kha-pa (2004). *The great treatise on the stages of the path to enlightenment.* Volume Two. Ithaca, NY: Snow Lion Publications.

Tsong-kha-pa (2005). *The six Yogas of Naropa.* Ithaca, NY: Snow Lion Publications.

Tsong-kha-pa (2006). *The harmony of emptiness and dependent-arising*. Dharmasala: The Library of Tibetan Works and Archives.

Tumposky, N.R. (1984). Behavioral objectives, the cult of efficiency, and foreign language learning: Are they compatible? *TESOL Quarterly*, Vol. *18*, No. 2 (Jun., 1984), pp. 295–310.

Turner, V. (1987). Betwixt and between: The liminal period in rites of passage. In L. Mahdi, S. Foster, & M. Little, eds. *Betwixt and between: Patterns of masculine and feminine initiation*. La Salle: Open Court, pp. 3–41.

Tyler, R.W. (1949). *Basic principles of curriculum and instruction*. Chicago: The University of Chicago Press.

Wallace, B. (1987). *The stubborn particulars of grace*. Toronto: McClelland and Stewart.

Waring, C. (2011). Smoke and mirrors: A stunning visual epitaph to jazz's creates every picture taker. In *Mojo Music Magazine*™. Issue 206, January 2011, p. 120.

Warren, J. (2012). Whales are people too. *Readers' Digest Canada*. (July 2012). Retrieved from: http://www.readersdigest.ca/?q=magazine/true-stories/derrid-whales-are-people-too.

Watts, S. (2006). *The peoples' tycoon: Henry Ford and the American century*. New York: Vintage Books.

Webb, P. (2002). Peek-a-boo. Phyllis Webb interviewed by Jay Ruzesky. In T. Bowling (ed.), *Where the words come from: Canadian poets in conversation*. Roberts Creek, BC: Nightwood Editions.

Weinsheimer, J. (1985). *Gadamer's hermeneutics*. New Haven, CT: Yale University Press.

Whyte, D. (1998). *Entering the house of belonging*. Langley, WA: Many Rivers Company.

Whyte, D. (2001). *Crossing the unknown sea: Work as a pilgrimage of identity*. New York: Riverhead Books.

Wiebe, R. & Johnson, Y. (1998). *Stolen Lives: The Journey of a Cree Woman*. Toronto: Alfred Knopf Canada.

Wilson, B. & Parks, V.D. (1966). "Wonderful". Lyrics by V.D. Parks, © Warner/Chappell Music, Inc., Universal Music Publishing Group.

Wittgenstein, L. (1968). *Philosophical investigations*. Cambridge, UK: Blackwell.

Wrege, C. D. & Greenwood, R. (1991). *Frederick W. Taylor: The father of scientific management: Myth and reality*. New York: Irwin Professional Publishing. Currently out of print. The text of Chapter 9 is available on-line at: johntaylorgatto.com/chapters/9d.hFtm.

Yangsi Rinpoche (2003). *Practicing the path: A commentary on the Lamrim Chenmo*. Boston: Wisdom Publications.

Yolen, J. (1988). *Favorite folktales from around the world*. New York: Pantheon Books.

Yolen, J. (1998). *Here there be dragons*. London, UK: Sandpiper Books.

Index

Y

Z

Studies in the Postmodern Theory of Education

General Editor
Shirley R. Steinberg

Counterpoints publishes the most compelling and imaginative books being written in education today. Grounded on the theoretical advances in criticalism, feminism, and postmodernism in the last two decades of the twentieth century, Counterpoints engages the meaning of these innovations in various forms of educational expression. Committed to the proposition that theoretical literature should be accessible to a variety of audiences, the series insists that its authors avoid esoteric and jargonistic languages that transform educational scholarship into an elite discourse for the initiated. Scholarly work matters only to the degree it affects consciousness and practice at multiple sites. Counterpoints' editorial policy is based on these principles and the ability of scholars to break new ground, to open new conversations, to go where educators have never gone before.

For additional information about this series or for the submission of manuscripts, please contact:

Shirley R. Steinberg
c/o Peter Lang Publishing, Inc.
29 Broadway, 18th floor
New York, New York 10006

To order other books in this series, please contact our Customer Service Department:

(800) 770-LANG (within the U.S.)
(212) 647-7706 (outside the U.S.)
(212) 647-7707 FAX

Or browse online by series:

www.peterlang.com